Regression Analysis Using MATLAB

Practical Techniques, Real Data Examples, and Step-by-Step Modeling

Peter I. Kattan

BOOKBOUND PRESS
Book Bound Press
https://bookboundpress.gumroad.com/

Peter I. Kattan, PhD

Correspondence about this book may be sent to the author at one of the following two email addresses:

pkattan@PetraBooks.com

info@PetraBooks.com

Regression Analysis Using MATLAB, written by Peter I. Kattan.

ISBN: 979-8-2955-4914-4

In Loving Memory of My Parents

Issa and Mary

Preface

Regression analysis remains one of the most powerful and widely used tools in statistics, data science, and scientific modeling. Whether predicting future trends, quantifying relationships between variables, or building mathematical models of natural and engineered systems, regression provides the analytical backbone for evidence-based decision-making. MATLAB, with its rich set of numerical and visualization capabilities, offers an ideal environment for learning, applying, and mastering regression techniques.

This book—*Regression Analysis Using MATLAB*—evolves from my earlier text on MATLAB and statistics, which introduced readers to foundational MATLAB operations and basic statistical concepts. In response to the feedback of students, instructors, and researchers, this new edition focuses exclusively on regression analysis, offering deeper coverage, expanded examples, and more comprehensive modeling strategies. Here, the reader will find intuitive explanations, step-by-step MATLAB implementations, and practical demonstrations using real and synthetic data.

The material progresses from simple linear regression to multiple regression, and polynomial models. Special attention is given to MATLAB's built-in functions, and the logical workflow of constructing, validating, and applying regression models in scientific research and applied fields.

My hope is that this book will serve as a practical resource for students in engineering, science, business, and the social sciences, as well as researchers who rely on quantitative modeling in their work. As with my previous publications, I welcome comments and suggestions for future editions.

January 2026 Peter I. Kattan

Table of Contents

Chapter 1

Introduction

In this introductory chapter a short MATLAB tutorial is provided. This tutorial describes the basic MATLAB commands needed. More details about these commands will follow in subsequent chapters. Note that there are numerous free MATLAB tutorials on the internet – check references [1-28]. Also, you may want to consult many of the excellent books on the subject – check references [29-47]. This chapter may be skipped on a first reading of the book.

In this tutorial it is assumed that you have started MATLAB on your computer system successfully and that you are now ready to type the commands at the MATLAB prompt (which is denoted by double arrows ">>"). For installing MATLAB on your computer system, check the web links provided at the end of the book.

Entering scalars and simple operations is easy as is shown in the examples below:

```
>> 2*3+5

ans =

    11
```

The order of the operations will be discussed in subsequent chapters.

```
>> cos(60*pi/180)

ans =

    0.5000
```

The argument for the `cos` command and the value of `pi` will be discussed in subsequent chapters.

We assign the value of 4 to the variable `x` as follows:

```
>> x = 4

x =

     4

>> 3/sqrt(2+x)

ans =

    1.2247
```

To suppress the output in MATLAB use a semicolon to end the command line as in the following examples. If the semicolon is not used then the output will be shown by MATLAB:

```
>> y = 30;
>> z = 8;
>> x = 2; y-z;
>> w = 4*y + 3*z
```

```
w =

    144
```

MATLAB is case-sensitive, i.e. variables with lowercase letters are different than variables with uppercase letters. Consider the following examples using the variables x and X:

```
>> x = 1

x =

      1

>> X = 2

X =

      2

>> x

x =

      1

>> X

X =

      2
```

Use the help command to obtain help on any particular MATLAB command. The following example demonstrates the use of `help` to obtain help on the `det` command which calculated the determinant of a matrix:

```
>> help det
 DET    Determinant.
    DET(X) is the determinant of
the square matrix X.

    Use COND instead of DET to
test for matrix singularity.

    See also cond.

    Overloaded functions or
methods (ones with the same name
in other directories)
       help sym/det.m

    Reference page in Help browser
       doc det
```

The following examples show how to enter matrices and perform some simple matrix operations:

```
>> x = [1 2 3 ; 4 5 6 ; 7 8 9]

x =

     1     2     3
     4     5     6
     7     8     9

>> y = [1 ; 0 ; -4]

y =
```

```
     1
     0
    -4

>> w = x*y

w =

   -11
   -20
   -29
```

Let us now see how to use MATLAB to solve a system of simultaneous algebraic equations. Let us solve the following system of simultaneous algebraic equations:

$$\begin{bmatrix} 3 & 5 & -1 \\ 0 & 4 & 2 \\ -2 & 1 & 5 \end{bmatrix} \begin{Bmatrix} x_1 \\ x_2 \\ x_3 \end{Bmatrix} = \begin{Bmatrix} 2 \\ 1 \\ -4 \end{Bmatrix} \quad (1.1)$$

We will use Gaussian elimination to solve the above system of equations. This is performed in MATLAB by using the backslash operator "\" as follows:

```
>> A= [3 5 -1 ; 0 4 2 ; -2 1 5]

A =

     3     5    -1
     0     4     2
    -2     1     5
```

```
>> b = [2 ; 1 ; -4]

b =

     2
     1
    -4

>> x = A\b

x =

   -1.7692
    1.1154
   -1.7308
```

It is clear that the solution is x_1 = -1.7692, x_2 = 1.1154, and x_3 = -1.7308. Alternatively, one can use the inverse matrix of A to obtain the same solution directly as follows:

```
>> x = inv(A)*b

x =

   -1.7692
    1.1154
   -1.7308
```

It should be noted that using the inverse method usually takes longer than using Gaussian elimination especially for large systems of equations.

Consider now the following 4 x 4 matrix D:

```
>> D = [1 2 3 4 ; 2 4 6 8 ; 3 6 9
12 ; -5 -3 -1 0]

D =

     1     2     3     4
     2     4     6     8
     3     6     9    12
    -5    -3    -1     0
```

We can extract the sub-matrix in rows 2 to 4 and columns 1 to 3 as follows:

```
>> E = D(2:4, 1:3)

E =

     2     4     6
     3     6     9
    -5    -3    -1
```

We can extract the third column of D as follows:

```
>> F = D(1:4, 3)

F =

     3
     6
     9
    -1
```

We can extract the second row of D as follows:

```
>> G = D(2, 1:4)

G =

     2     4     6     8
```

We can also extract the element in row 3 and column 2 as follows:

```
>> H = D(3,2)

H =

     6
```

We will now show how to produce a two-dimensional plot using MATLAB. In order to plot a graph of the function $y = f(x)$, we use the MATLAB command `plot(x,y)` after we have adequately defined both vectors `x` and `y`. The following is a simple example to plot the function $y = x^2 + 3$ for a certain range:

```
>> x = [1 2 3 4 5 6 7 8 9 10 11]

x =

  Columns 1 through 10

     1          2          3          4          5
6     7     8     9     10

  Column 11

    11
```

```
>> y = x.^2 + 3

y =

  Columns 1 through 10

      4         7        12        19        28
39       52        67        84       103

  Column 11

    124

>> plot(x,y)
```

Figure 1.1 shows the plot obtained by MATLAB. It is usually shown in a separate window. In this figure no titles are given to the x- and y-axes. These titles may be easily added to the figure using the `xlabel` and `ylabel` commands. More details about these commands will be presented in the chapter on graphs.

In the calculation of the values of the vector y, notice the dot "." before the exponentiation symbol "^". This dot is used to denote that the operation following it be performed element by element. More details about this issue will be discussed in subsequent chapters.

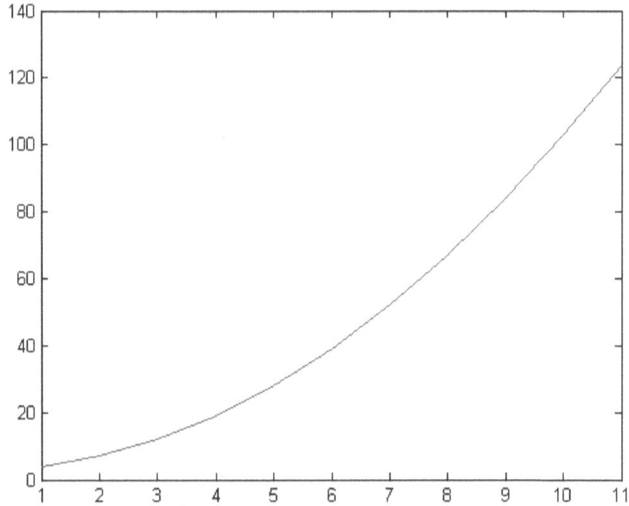

Figure 1.1: Using the MATLAB "plot" command

Finally, we will show how to make magic squares with MATLAB. A magic square is a square grid of numbers where the total of any row, column, or diagonal is the same. Magic squares are produced by MATLAB using the command `magic`. Here is a simple example to produce a 5 x 5 magic square:

```
>> magic(5)

ans =

    17    24     1     8    15
    23     5     7    14    16
     4     6    13    20    22
    10    12    19    21     3
    11    18    25     2     9
```

It is clear that the total of each row, column or diagonal in the above matrix is 65.

Exercises

Solve all the exercises using MATLAB. All the needed MATLAB commands for these exercises were presented in this chapter.

1. Perform the operation 3*4+6. The order of the operations will be discussed in subsequent chapters.

2. Perform the operation $\cos(5)$. The value of 5 is in radians.

3. Perform the operation $3\sqrt{6+x}$ for $x = 4$.

4. Assign the value of 5.2 to the variable y.

5. Assign the values of 3 and 4 to the variables x and y, respectively, then calculate the value of z where $z = 2x - 7y$.

6. Obtain help on the `inv` command.

7. Generate the following matrix: A

$$A = \begin{bmatrix} 1 & 0 & 2 & -3 \\ 0 & 5 & 2 & 2 \\ 1 & 2 & 3 & 4 \\ -2 & 0 & 1 & 3 \end{bmatrix}$$

8. Generate the following vector b

$$b = \begin{Bmatrix} 1 \\ 2 \\ 3 \\ 4 \end{Bmatrix}$$

9. Evaluate the vector c where $\{c\} = [A]\{b\}$ where A is the matrix given in Exercise 7 above and b is the vector given in Exercise 8 above.

10. Solve the following system of simultaneous algebraic equation using Gaussian elimination.

$$\begin{bmatrix} 5 & 2 \\ 1 & 3 \end{bmatrix} \begin{Bmatrix} x_1 \\ x_2 \end{Bmatrix} = \begin{Bmatrix} 3 \\ -1 \end{Bmatrix}$$

11. Solve the system of simultaneous algebraic equations of Exercise 10 above using matrix inversion.

12. Generate the following matrix X:

$$X = \begin{bmatrix} 1 & 0 & 6 \\ 1 & 2 & 3 \\ 4 & 5 & -2 \end{bmatrix}$$

13. Extract the sub-matrix in rows 2 to 3 and columns 1 to 2 of the matrix X in Exercise 12 above.

14. Extract the second column of the matrix X in Exercise 12 above.

15. Extract the first row of the matrix X in Exercise 12 above.

16. Extract the element in row 1 and column 3 of the matrix X in Exercise 12 above.

17. Generate the row vector x with integer values ranging from 1 to 9.

18. Plot the graph of the function $y = x^3 - 2$ for the range of the values of x in Exercise 17 above.

19. Generate a 4 x 4 magic square. What is the total of each row, column, and diagonal in this matrix.

Chapter 2

Statistics Review – Part I

In this chapter the basic numerical descriptive measures of statistical data are discussed. Measures of central tendency and measures of dispersion are presented including the mean, median, variance, and standard deviation. All the respective MATLAB commands[1] used for statistics in this chapter are illustrated with examples. There is no coverage of probability theory – only basic statistical concepts.

There are three measures of central tendency that are used most often. These are the arithmetic mean, median, and mode. Others are also discussed like the geometric mean and the harmonic mean. As for the measures of dispersion, the two most used measures are the variance and standard deviation. The percentiles and quartiles are also used but these two do not measure dispersion.

Let us start by listing the data for the example to be used throughout this chapter. The heights of 30 people in a neighborhood are measured in centimeters and are listed below:

118	147	158	162	111
150	162	155	143	170

[1] All the MATLAB commands used for statistics in this chapter are found in the main MATLAB package. No resort is made to the Statistics Toolbox that is purchased separately. The Statistics Toolbox includes specialized and advanced commands for Statistics but these are beyond the scope of this book.

192	181	151	144	137
171	180	190	163	165
157	148	188	182	170
170	181	179	182	160

The above data are stored in a vector in MATLAB. Let us call this vector `x`. The following is the relevant MATLAB command:

```
>> x = [118 147 158 162 111 150 162 155
143 170 192 181 151 144 137 171 180 190
163 165 157 148 188 182 170 170 181 179
182 160]

x =

  Columns 1 through 12

   118    147    158    162    111    150
162    155    143    170    192    181

  Columns 13 through 24

   151    144    137    171    180    190
163    165    157    148    188    182

  Columns 25 through 30

   170    170    181    179    182    160
```

Alternatively, and if the data set is too long, one my store it in a text file (let us call it file_name), and then import the text file into MATLAB using the command `load` as follows:

```
x = load(file_name)
```

Let us first determine the size of the data set, i.e. the number of data points stored in the vector `x`.

For this purpose, we use the MATLAB command `size` as follows:

```
>> size(x)

ans =

     1    30
```

Thus it is clear that the vector X has 30 data points. Next, let us sort the data in the vector X using the MATLAB command `sort` as follows:

```
>> sort(X)

ans =

  Columns 1 through 10

   111    118    137    143    144    147
148    150    151    155

  Columns 11 through 20

   157    158    160    162    162    163
165    170    170    170

  Columns 21 through 30

   171    179    180    181    181    182
182    188    190    192
```

The measures of central tendency all try to find a value from the data set that lies at the center of the data. Different ways are used to do that involving the arithmetic mean, geometric mean, harmonic mean, median, mode, etc.

The arithmetic mean[2] of the heights is obtained by using the MATLAB command `mean` as follows:

```
>> mean(x)

ans =

  162.2333
```

The mean of a data set consisting of a number of data points is the arithmetic average of the data set. The mean is usually denoted by the symbol $\overline{x}$ for a data set consisting of n points x_i , $i = 1,2,3,.....,n$. The mean is usually calculated by hand using the following formula (however the use of the simple formula below is left to the reader; is suffices to use the MATLAB command as shown above):

$$\overline{x} = \frac{\sum_{i=1}^{n} x_i}{n}$$

The median of a data set consisting of a number of data points is literally the middle value when the data points are sorted or ordered. In this case, half the values are above the median and the other half is below the median. The median of the heights is obtained by using the MATLAB command `median` as follows:

```
>> median(x)
```

---

[2] Two other types of "mean" are also available in MATLAB. The geometric mean and the harmonic mean are calculated using the MATLAB commands geomean and harmean, respectively, but these are beyond the scope of this book.

```
ans =

  162.5000
```

The mode of the heights is obtained by using the MATLAB command `mode` as follows:

```
>> mode(x)

ans =

   170
```

It is noted that if the data is symmetrical, i.e. if it is symmetrically distributed, then the values of the arithmetic mean, the median, and the mode will be identical.

The measures of dispersion are values not from the data set, but are calculated through different formulas that try to show how the data differ or deviate from the central value. Such measures include the average deviation, variance, standard deviation, etc.

The range is simply the difference between the largest data value and smallest data value. The range of the give data is obtained using MATLAB using the command `range` as follows:

```
>> range(x)

ans =

    81
```

The interquartile range is the difference between the third quartile value and the first quartile value. A quartile (or a quarter) is one fourth of the data. Thus, the first quartile is the data value that has 25% of the data below it, while the third quartile is the data value that has 75$ of the data below it. Actually, the interquartile range is a measure of the middle 50% of the data. The interquartile range (IQR) of the give data is obtained using MATLAB using the command `iqr` as follows:

```
>> iqr(x)

ans =

    30
```

The variance s^2 of the data is a measure of the spread of the data. The variance is calculated by hand using the following formula:

$$s^2 = \frac{\sum_{i=1}^{n}(x_i - \bar{x})^2}{n-1}$$

The variance of the data is obtained by using the MATLAB command `var` as follows:

```
>> var(x)

ans =

  393.2195
```

The standard deviation s is also a measure of the spread of the data but is much more useful than the variance. The standard deviation is calculated manually using the following formula:

$$s^2 = \sqrt{\frac{\sum_{i=1}^{n}(x_i - \bar{x})^2}{n-1}}$$

The standard deviation of the data is obtained by using the MATLAB command `std` as follows:

```
>> std(x)

ans =

   19.8298
```

In all the above calculation, it is assumed that the data set of heights is a sample. There is no mention of the population, but it is noted that there are variations of the above measures that apply to the population, instead of the sample. Such measures would include the population mean and the population standard deviation, but these are beyond the scope of this work.

Next, we disucss measures of position, basically the quartiles, the percentiles, and the standardized scores. Quartiles divide the data into four equal parts, each of which contains 25% of the data. A percentile is one of 99 values that divide a set of data such that it gives the percentage of data values that lie below the given value. The `pth` percentile P is defined as the value of the data such that p percent of the data are less than P and (100-p) percent of the data are greater than P. The three most popular `pth` percentiles are the `25th` percentile (called the first quartile), the `50th` percentile (called the second or middle quartile), and the `75th` percentile (called the third quartile).

The following are some examples using MATLAB showing how use the different commands to calcualate some percentiles and quartiles for the data set used in this chapter:

The 0.30 quantile is calculated using MATLAB using the command `quantile` as follows:

```
>> quantile(x,0.30)

ans =

   153
```

The 0.50 quantile is calculated using MATLAB using the command `quantile` as follows:

```
>> quantile(x,0.50)

ans =

  162.5000
```

The 0.70 quantile is calculated using MATLAB using the command `quantile` as follows:

```
>> quantile(x,0.70)

ans =

   175
```

The 42nd percentile is calculated using MATLAB using the command `prctile` as follows:

```
>> prctile(x,42)
```

```
ans =

  160.2000
```

The 65th percentile is calculated using MATLAB using the command `prctile` as follows:

```
>> prctile(x,65)

ans =

   170
```

The 87th percentile is calculated using MATLAB using the command `prctile` as follows:

```
>> prctile(x,87)

ans =

   182
```

Finally, standardized scores (zscores) are used as a measue to to give the position of a data value relative to its mean and standard deviation. The zscore z is calculated by hand using the following formula:

$$z = \frac{x - \overline{x}}{s}$$

Based on the above formula, it is noted that for each data value x, a corresponding zscore value z is calculated. Thus, for a data set consisting of n data points, one can calculate n different zscores. The manual use of the above formula is left to the reader but calculating zscores using MATLAB will be illustrated below.

Standardized scores (also called zscores) can be calculated using MATLAB using the command `zscore` as follows:

```
>> zscore(x)

ans =

  Columns 1 through 7

   -2.2307    -0.7682    -0.2135    -0.0118
-2.5837    -0.6169    -0.0118

  Columns 8 through 14

   -0.3648    -0.9699     0.3917     1.5011
0.9464    -0.5665    -0.9195

  Columns 15 through 21

   -1.2725     0.4421     0.8960     1.4003
0.0387     0.1395    -0.2639

  Columns 22 through 28

   -0.7178     1.2994     0.9968     0.3917
0.3917     0.9464     0.8455

  Columns 29 through 30

     .9968      -0.1126
```

Other descriptive statistical measures used are the maximum value and minimum value of the data set. These three measures are illustrated below by using the MATLAB commands `max` and `min` as follows:

```
>> max(x)
```

```
ans =

   192

>> min(x)

ans =

   111
```

Finally, additional statistical measures can be calculated such as skewness, kurtosis, and moment using the MATLAB commands `skewness`, `kurtosis`, and `moment`, respectively. But these commands will not be illustrated here as they are beyond the scope of this beginner's book.

In the remainder of this chapter, we discuss how to calculate Pearson's correlation coefficient[3]. This linear correlation coefficient is used to meaure the strength of a linear relationship between two variables. This measurement has limits at -1 and +1. The endpoints indicate a perfect correlation. The farther the measurement is from zero, the stronger the relationship between the two variables.

The Pearson correlation coefficient is illustrated in the next example. For this purpose, consider two sets of data or observations that need to be compared. Each set of data describes an independent variable. The value of the correlation coefficient lies between +1 and -1 inclusive. Total positive correlation exists when the value of the correlation coefficient is equal to +1. No correlation

---

[3] Other types of correlation coefficients exist in the statistics literature, but these are beyond the scope of this book.

exists when the value of the correlation coefficient is equal to 0. Total negative correlation exists when the value of the correlation coefficient is equal to -1. If the value of the correlation coefficient is positive, the positive correlation exists between the two variables. On the other hand, negative correlation exists between the two variables if the value of the correlation coefficient is negative. The two data sets for this example are shown below.

Set A Data:

11 14 17 20 24 56
60
66 70 72 77 81 90
100
101 110 116 118 120 122
140

Set B Data:

1.4 2.5 2.6 3.2 3.3 3.5
4.1
4.5 4.7 5.0 5.6 6.0 6.6
7.0
7.8 8.0 9.0 9.9 10.1 11.1
12.0

In order to conduct a correlation analysis between the two sets of data A and B, use the MATLAB command `corrcoef` to calculate the Pearson correlation coefficient of these two sets of data as follows:

```
>> x = [11 14 17 20 24 56 60 66 70 72 77
81 90 100 101 110 116 118 120 122 140]
```

```
x =

  Columns 1 through 12

    11       14       17       20       24       56
60     66     70     72     77     81

  Columns 13 through 21

    90      100      101      110      116      118
120    122    140

>> y = [1.4 2.5 2.6 3.2 3.3 3.5 4.1 4.5
4.7 5.0 5.6 6.0 6.6 7.0 7.8 8.0 9.0 9.9
10.1 11.1 12.0]

y =

  Columns 1 through 7

    1.4000      2.5000      2.6000      3.2000
3.3000     3.5000     4.1000

  Columns 8 through 14

    4.5000      4.7000      5.0000      5.6000
6.0000     6.6000     7.0000

  Columns 15 through 21

    7.8000      8.0000      9.0000      9.9000
10.1000    11.1000    12.0000

>> corrcoef(x,y)

ans =

    1.0000    0.9576
    0.9576    1.0000
```

Thus, the value of the correlation coefficient in this example is 0.9576 indicating a strong positive correlation. Alternatively, instead of using the MATLAB command `corrcoef` directly, one can use the formula for Pearson's correlation coefficient, and calculate the value of the coefficient by hand. The formula is given by the following equation for two sets x_i and y_i of n data points:

$$r = \frac{n\sum_{i=1}^{n} x_i y_i - \sum_{i=1}^{n} x_i \sum_{i=1}^{n} y_i}{\sqrt{n\sum_{i=1}^{n} x_i^2 - \left(\sum_{i=1}^{n} x_i\right)^2}\sqrt{n\sum_{i=1}^{n} y_i^2 - \left(\sum_{i=1}^{n} y_i\right)^2}}$$

As an example of using MATLAB, we will illustrate how to use the above formula in order to calculate the value of the correlation coefficient a second time. In order to use MATLAB to compute the above formula for the two data sets in the example above, one needs first to calculate the sums $\sum_{i=1}^{n} x_i$, $\sum_{i=1}^{n} y_i$, $\sum_{i=1}^{n} x_i^2$, $\sum_{i=1}^{n} y_i^2$, and $\sum_{i=1}^{n} x_i y_i$. For this purpose, Table 12.1 shows each sum along with the required MATLAB command associated with it:

Table 12.1: MATLAB Command Required for Each Respective Sum

Sum	MATLAB Command
x_i	`x`
y_i	`y`
$\sum_{i=1}^{n} x_i$	`sum(x)`

$\sum_{i=1}^{n} y_i$	`sum(y)`
$\sum_{i=1}^{n} x_i^2$	`mxx=x.*x` `sum(sxx)`
$\sum_{i=1}^{n} y_i^2$	`myy=y.*y` `sum(syy)`
$\sum_{i=1}^{n} x_i y_i$	`mxy=x.*y` `sum(sxy)`

The following are the MATLAB commands needed to calculate Pearson's correlation coefficient for the example above using the formula given above:

Step One: Enter the data points given in the Example:

```
>> x = [11 14 17 20 24 56 60 66 70 72 77
81 90 100 101 110 116 118 120 122 140]

x =

  Columns 1 through 12

    11      14      17      20      24      56
60      66      70      72      77      81

  Columns 13 through 21

    90     100     101     110     116     118
120    122    140

>> y = [1.4 2.5 2.6 3.2 3.3 3.5 4.1 4.5
4.7 5.0 5.6 6.0 6.6 7.0 7.8 8.0 9.0 9.9
10.1 11.1 12.0]

y =
```

```
  Columns 1 through 7

    1.4000      2.5000      2.6000      3.2000
3.3000      3.5000      4.1000

  Columns 8 through 14

    4.5000      4.7000      5.0000      5.6000
6.0000      6.6000      7.0000

  Columns 15 through 21

    7.8000      8.0000      9.0000      9.9000
10.1000     11.1000     12.0000
```

Step Two: Calculate the five different sums needed to compute the coefficient using the formula:

2.1 Calcuate the sum $\sum_{i=1}^{n} x_i$:

```
>> sumx = sum(x)

sumx =

        1585
```

2.2 Calculate the sum $\sum_{i=1}^{n} y_i$:

```
>> sumy = sum(y)

sumy =

  127.9000
```

2.3: Calculate the sum $\sum_{i=1}^{n} x_i^2$:

```
>> mxx = x.*x

mxx =

  Columns 1 through 5

         121              196              289
400          576

  Columns 6 through 10

        3136             3600             4356
4900         5184

  Columns 11 through 15

        5929             6561             8100
10000        10201

  Columns 16 through 20

       12100            13456            13924
14400        14884

  Column 21

       19600

>> sumxx = sum(mxx)

sumxx =

      151913
```

2.4: Calculate the sum $\sum_{i=1}^{n} y_i^2$:

```
>> myy = y.*y

myy =
```

```
  Columns 1 through 7

    1.9600     6.2500     6.7600    10.2400
10.8900   12.2500   16.8100

  Columns 8 through 14

   20.2500    22.0900    25.0000    31.3600
36.0000   43.5600   49.0000

  Columns 15 through 21

   60.8400    64.0000    81.0000    98.0100
102.0100  123.2100  144.0000

>> sumyy = sum(myy)

sumyy =

  965.4900
```

2.5: Calculate the sum $\sum_{i=1}^{n} x_i y_i$:

```
>> mxy = x.*y

mxy =

   1.0e+03 *

  Columns 1 through 7

    0.0154     0.0350     0.0442     0.0640
0.0792    0.1960    0.2460

  Columns 8 through 14

    0.2970     0.3290     0.3600     0.4312
0.4860    0.5940    0.7000

  Columns 15 through 21
```

```
    0.7878      0.8800      1.0440      1.1682
1.2120     1.3542     1.6800

>> sumxy = sum(mxy)

sumxy =

   1.2003e+04
```

2.6: Calculate the square $\left(\sum_{i=1}^{n} x_i\right)^2$:

```
>> sumx2 = sumx*sumx

sumx2 =

     2512225
```

2.7: Calculate the square $\left(\sum_{i=1}^{n} y_i\right)^2$:

```
>> sumy2 = sumy*sumy

sumy2 =

   1.6358e+04
```

2.8: Calculate the number of data points in the data set using the `size` command:

```
>> n = size(x)

n =

     1    21

>> n1 = n(2)

n1 =
```

```
    21
```

Step Three: Use the formula to calculate the value of the correlation coefficient by substituting the values of the sums obtained above:

```
>>    r    =    (n1*sumxy    -    sumx*sumy)    /
(sqrt(n1*sumxx-sumx2)*sqrt(n1*sumyy-
sumy2))

r =

    0.9576
```

It is clear from the result above that the correlation between the two data sets is positive and high. The value of the coefficient is very close to 1.0. It is noted that the exact same value of the correlation coefficient (i.e. 0.9576) is obtained also using the MATLAB command corrcoef as was shown previously (at the beginning of this example).

Exercises

Solve Exercises 1 - 8 using the following set of 25 data points using MATLAB:

25	18	17	24	11
10	20	30	25	12
11	12	15	28	33
20	12	20	22	30
15	18	22	19	31

1. Store the data set in a vector called x using MATLAB.

2. Sort the data set using MATLAB and find its size (i.e. the number of data points) using MATLAB.

3. Calculate the mean, median and mode for the data set using MATLAB.

4. Calculate the range, the interquantile range, the variance, and the standard deviation for the data set using MATLAB.

5. Calculate the maximum and minimum values using MATLAB.

6. Calculate the 0.30 quantile, the 0.50 quantile, and the 0.70 quantile for this data set.

7. Calculate the 47nd percentile, the 55th percentile, and the 80th percentile for this data set.

8. Calculate the standardized scores (zscore) for this data set.

Solve Exercies 9-11 using the following two sets of data points using MATLAB:

Data Set One:

1.24 2.56 3.45 4.66 3.56
4.90
3.33 2.45 2.34 2.00 1.45
1.56

Data Set Two:

600 450 320 300 430
330
470 570 340 200 250
390

9. Calculate Pearson's correlation coefficient for the two sets of data points given above directly in MATLAB using the command `corrcoef`.

10. Calculate Pearson's correlation coefficient for the aboe two sets of data points using MATLAB but indirectly by using the formula given in this chapter. Note that you need to calculate the various sums appearing in the formula first.

11. Compare the two results that you obtained for Pearson's correlation coefficient from Exercises # 9 and # 10 above.

Chapter 3

Statistics Review – Part II

In this chapter the basic methods of graphical visualization of statistical data are discussed. Several types of graphs designed for statistical data are presented such as histograms, frequency diagrams, scatter diagram, and pie charts. All the respective MATLAB commands[4] used for statistics in this chapter are illustrated with examples.

In order to demonstrate the various graphical measures used in statistics, let us start by again listing the data for the same example of Chapter 12 to be used throughout this chapter. The heights of 30 people in a neighborhood are measured in centimeters and are listed below:

118	147	158	162	111
150	162	155	143	170
192	181	151	144	137
171	180	190	163	165
157	148	188	182	170
170	181	179	182	160

[4] All the MATLAB commands used for statistics in this chapter are found in the main MATLAB package and no resort is made to the Statistics Toolbox that is purchased separately. The Statistics Toolbox includes more specialized and advanced commands for Statistics but these are beyond the scope of this book.

The above data are stored in a vector in MATLAB. Let us call this vector x. The following is the relevant MATLAB command (repeated from Chapter 12):

```
>> x = [118 147 158 162 111 150 162 155
143 170 192 181 151 144 137 171 180 190
163 165 157 148 188 182 170 170 181 179
182 160]

x =

  Columns 1 through 12

   118      147      158      162      111      150
162    155    143    170    192    181

  Columns 13 through 24

   151      144      137      171      180      190
163    165    157    148    188    182

  Columns 25 through 30

   170    170    181    179    182    160
```

The bar chart (also called bar graph) of the data is obtained using the MATLAB command bar as follows. The resulting graph is shown in Figure 13.1. The bar chart shows the counts for each category of heights. Each category of information is depicted in a separate bar. Individual bars are separated by a space. The height of each bar represent the count or number of heights for that specific category.

```
>> bar(x)
>> hold on
>> xlabel('heights')
>> ylabel('Number of People')
>> xlabel('Heights')
```

The histogram of the data is obtained using the MATLAB command `hist` as follows. The resulting graph is shown in Figure 13.2. Like the bar chart, the histogram plots data in bars, where the height of the bar represents the count or number of heights. There are no spaces between bars in a histogram like those in bar charts.

```
>> hist(x)
>> hold on
>> xlabel('Heights')
>> ylabel('Number of People')
```

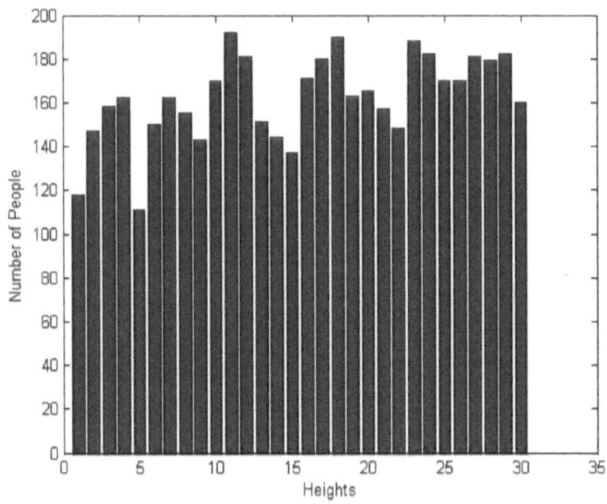

Figure 13.1: Bar Chart for the Example

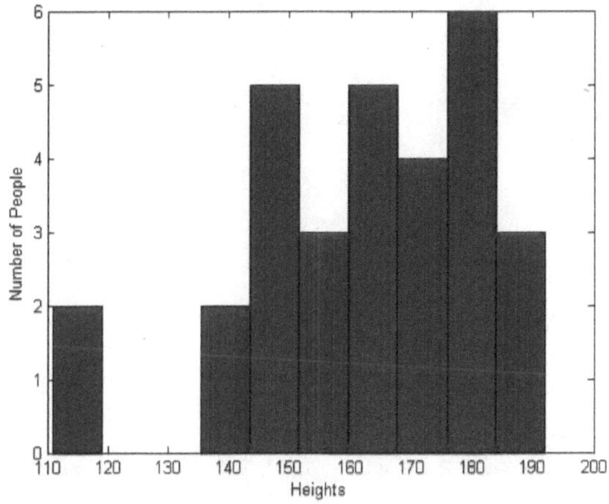

Figure 13.2: Histogram for the Example

The cumulative frequence plot of the data is obtained using the MATLAB command `cdfplot` as follows. The resulting graph is shown in Figure 13.3. The cumulative frequency plot is another type of histogram. The height of the first bar represents the proportion of data points that fall into the given value. The height of the second bar is the frequency or proportion of data points that fall into that given data plus those that fall into the first bar. Thus, any given bar represents the frequency or proportion of data values that fall into that value or anything less than that value.

```
>> cdfplot(x)
```

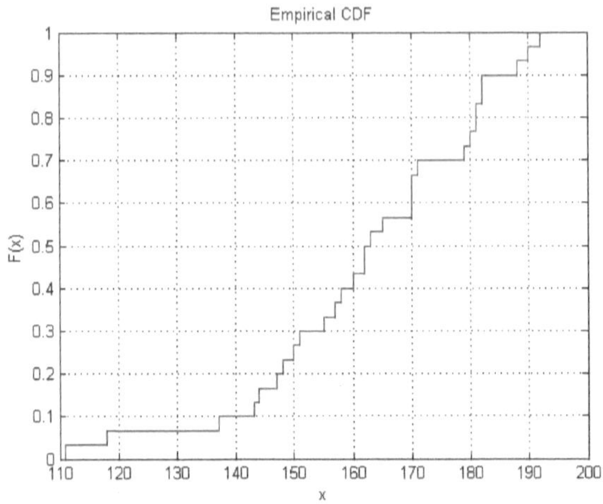

Figure 13.3: Cumulative Frequencey Plot for the Example

The scatterplot of the data is obtained using the MATLAB command `scatter` as follows. The resulting graph is shown in Figure 13.4. But first we need to store the number of data points in a variable (called `y` here) in order to plot them on the x-axis while the actual data points are plotted on the y-axis. Note that the scatterplot shows the heights (or data values) on the y-axis as points on the x-y graph.

```
>> size(x)

ans =

     1    30

>> y = 1:30:1

y =

     1

>> y = 1:30
```

```
y =

  Columns 1 through 11

     1          2          3          4          5          6
7      8        9       10       11

  Columns 12 through 22

    12         13         14         15         16         17
18      19       20       21       22

  Columns 23 through 30

    23         24         25         26         27         28
29      30

>> scatter(y,x)
>> hold on
>> ylabel('Heights')
>> xlabel('Number')
>> scatter(y,x)
```

The pie chart of the data is obtained using the MATLAB command pie as follows. The resulting graph is shown in Figure 13.5. Pie charts show the whole group or population as a circle or pie. The slices of the pie represent each category or subgroup. The sizes of the slices are proportional to the fraction of the whole that each category or subgroup represents. In this example, no attempt is made to group the data in categories or subgroups. The data is plotted into a pie chart as they are given.

```
>> pie(x)
```

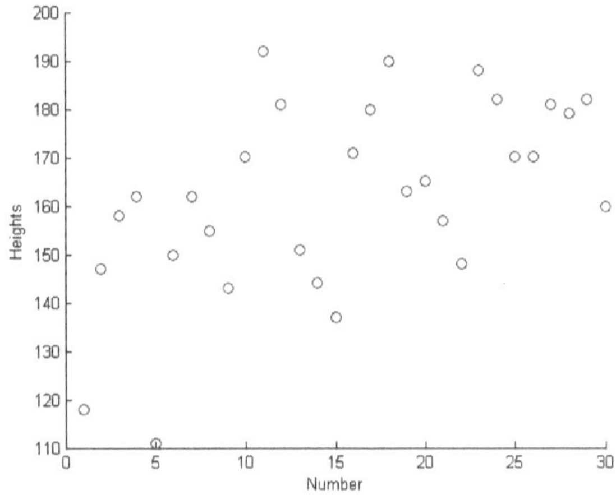

Figure 13.4: Scatterplot of the Data Points of the Example

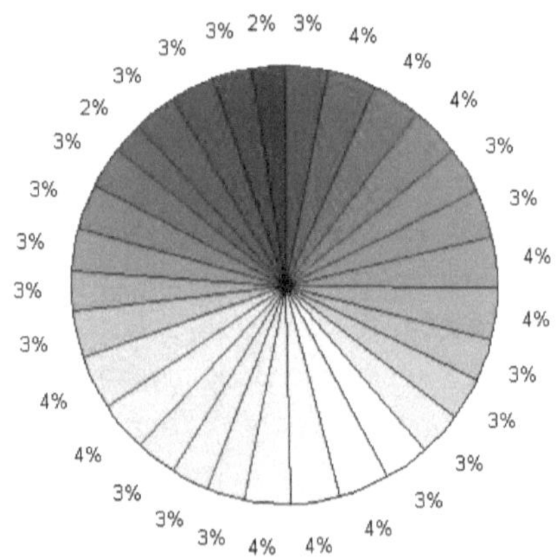

Figure 13.5: Pie Chart of the Data Points of the Example

The rose diagram (some sort of an angular histogram) of the data is obtained using the MATLAB command rose as follows. The resulting graph is shown in Figure 13.6. The rose diagram is circular like the pie chart but shows angular bars instead of slices.

```
>> rose(x)
```

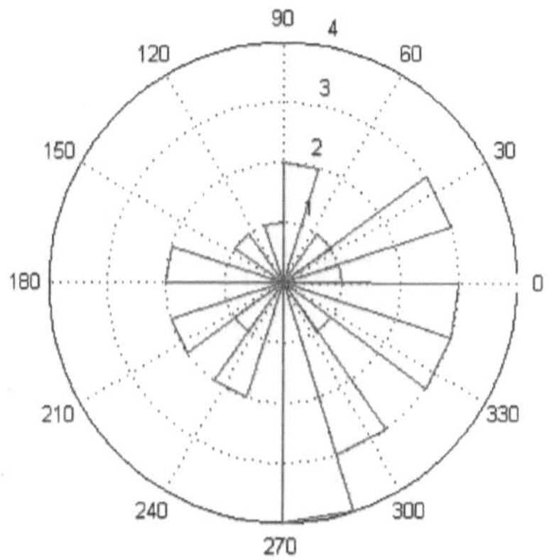

Figure 13.6: Rose Diagram for the Example

A stem plot of the data is obtained using the MATLAB command stem as follows. The resulting graph is shown in Figure 13.7. A stem plot is like a histogram or bar chart, but uses stems (vertical lines) instead of bars.

```
>> stem(x)
>> hold on
>> xlabel('Heights')
>> ylabel('Number of People')
```

A scatterplot of the zscores can also be plotted. Calculation of the zscores was shown previously in Chapter 12. This is repeated here and followed by using the MATLAB command scatter to plot the scatterplot of the zscores. The detailed

MATLAB commands follow. The resulting graph is shown in Figure 13.8.

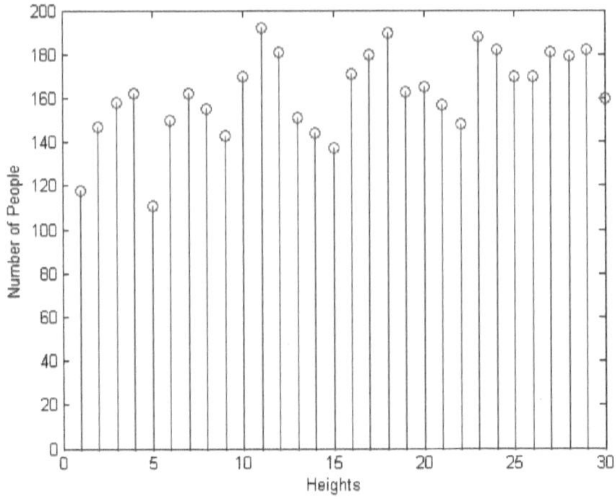

Figure 13.7: A Stem Plot of the Data Points of the Example

```
>> zscore(x)

ans =

  Columns 1 through 7

   -2.2307    -0.7682    -0.2135    -0.0118
-2.5837    -0.6169    -0.0118

  Columns 8 through 14

   -0.3648    -0.9699     0.3917     1.5011
0.9464    -0.5665    -0.9195

  Columns 15 through 21

   -1.2725     0.4421     0.8960     1.4003
0.0387     0.1395    -0.2639

  Columns 22 through 28
```

```
   -0.7178      1.2994      0.9968      0.3917
0.3917     0.9464     0.8455

  Columns 29 through 30

    0.9968    -0.1126

>> scatter(y,ans)
>> hold on
>> xlabel('Number')
>> ylabel('zscore')
```

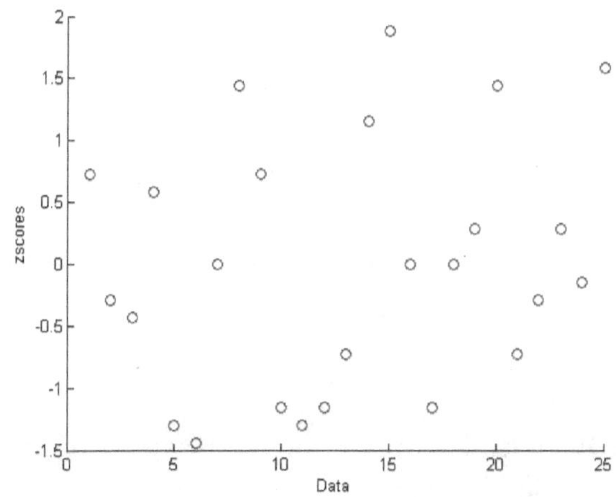

Figure 13.8: Scatterplot of Zscores for the Example

A boxplot (also called a box or whiskers plot) of the data points can also be plotted. This shown in MATLAB below using the command boxplot. The resulting graph is shown in Figure 13.9. A box plot graphically shows the distribution of data by focusing on five key measures: the minimum value, the first quartile(25th percentile), the median (50th percentile), the third quartile (75th percentile), and the maximum value.

```
>> boxplot(x)
>> hold on
>> ylabel('Heights')
```

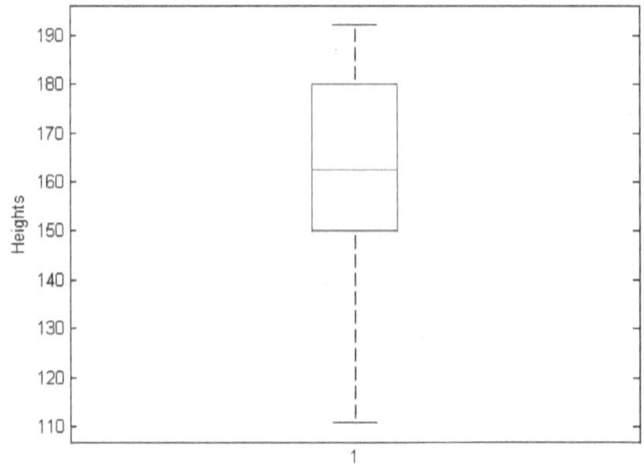

Figure 13.9: A Boxplot of the Data Points of the Example

Exercises

Solve Exercies 1 - 9 using the following set of 25 data points using MATLAB:

25	18	17	24	11
10	20	30	25	12
11	12	15	28	33
20	12	20	22	30
15	18	22	19	31

1. Draw a bar chart based on the above set of data points using MALTAB.

2. Draw a histogram based on the above set of data points using MALTAB.

3. Draw a cumulative frequency plot based on the above set of data points using MATLAB.

4. Draw a scatterplot using the data points in the Exercise above.

5. Draw a pie chart based on the data set in the Exercise above.

6. Draw a rose diagram based on the data set in the Exercise above.

7. Draw a stem plot based on the data points of the Exercise above.

8. Draw a scatterplot of the zscores of the data points of the Exercise above.

9. Draw a boxplot for the Exercise above.

Chapter 4

Regression Analysis – Part I

Regression analysis is a statistical method which helps to determine an approximate form of the relationship between variables. In this chapter we consider the relationship between two variables using simple linear regression[5]. The relationship between three or more variables is presented in Chapter 15 using multiple regression.

Two variables are related linearly if their relationship can be expressed by the equation of a straight line. This is the essence of simple linear regression. The resulting straight line is considered to be a best-fit for the data under study. We assume the equation of the straight line to be written in the following form:

$$y = a + bx$$

where a and b are constant coefficients to be determined for a specific set of data (the coefficient b represents the slope or gradient of the resulting straight line while the coefficient a represents the y-

[5] There are some specialized MATLAB commands like the command polyfit that can be used to find the best fitting polynomial or straight line for a give set of data. But such specialized and advanced commands are not used in this book. Instead use is made of the various MATLAB commands found in the basic package to solve these types of problems. In addition, using simple basic commands from scratch enhances the learning process considerably.

intercept, i.e. the point where the resulting straight line intersects the y-axis). We will proceed with the detailed derivation of the equations of simple linear regression.

For a set of data x_i and y_i (where $i = 1,2,3,......n$), the above equation becomes for each point (x_i, y_i):

$$y_i = a + bx_i$$
(A)

It should be noted that the above equations is repeated n times, where n is the number of data points. Let us write the above equations for the first few data points as $i = 1,2,3,......n$:

$$y_1 = a + bx_1$$

$$y_2 = a + bx_2$$

$$y_3 = a + bx_3$$

....

....

....

$$y_n = a + bx_n$$

Let us now add the above list of equations to obtain the following expression:

$$y_1 + y_2 + y_3 + \ldots\ldots + y_n = a + bx_1 + a + bx_2 + a + bx_3 + \ldots\ldots + a + bx_n$$

The above equation can be written in the following simplified form:

$$y_1 + y_2 + y_3 + \ldots\ldots + y_n = (a + a + a + \ldots. + a) + (bx_1 + bx_2 + bx_3 + \ldots\ldots + bx_n)$$

where the sum of the a's in the first parenthesis is n times. The above equation can be written using the summation sign $\sum$ as follows:

$$\sum_{i=1}^{n} y_i = na + b\sum_{i=1}^{n} x_i$$
(B)

where the following summation notation is used in equation (B):

$$\sum_{i=1}^{n} y_i = y_1 + y_2 + y_3 + \ldots\ldots\ldots + y_n$$

$$na = a + a + a + \ldots\ldots + a \qquad (summed\ n\ times)$$

$$b\sum_{i=1}^{n} x_i = b(x_1 + x_2 + x_3 + \ldots\ldots + x_n) = bx_1 + bx_2 + bx_3 + \ldots. + bx_n$$

The next step would be to multiply equation (A) by x_i to obtain the following equation:

$$x_i y_i = a x_i + b x_i x_i$$
(C)

Writing the above equation out for each value $i = 1, 2, 3, \ldots\ldots n$, we obtain:

$$x_1 y_1 = a x_1 + b x_1^2$$

$$x_2 y_2 = a x_2 + b x_2^2$$

$$x_3 y_3 = a x_3 + b x_3^2$$

....

....

....

$$x_n y_n = a x_n + b x_n^2$$

Adding the above list of equations results in the following expression:

$$x_1 y_1 + x_2 y_2 + x_3 y_3 + \ldots\ldots + x_n y_n = a x_1 + b x_1^2 + a x_2 + b x_2^2 + a x_3 + b x_3^2 + \ldots\ldots\ldots + a x_n + b x_n^2$$

The above equation can be simplified as follows:

$$x_1 y_1 + x_2 y_2 + x_3 y_3 + \ldots\ldots + x_n y_n = a x_1 + a x_2 + a x_3 + \ldots\ldots + a x_n + b x_1^2 + b x_2^2 + b x_3^2 + \ldots\ldots + b x_n^2$$

The above equation can be written using the summation sign $\sum$ as follows:

$$\sum_{i=1}^{n} x_i y_i = a\sum_{i=1}^{n} x_i + b\sum_{i=1}^{n} x_i^2$$

(D)

where the following summation notation is used in equation (D):

$$\sum_{i=1}^{n} x_i y_i = x_1 y_1 + x_2 y_2 + x_3 y_3 + ... + x_n y_n$$

$$a\sum_{i=1}^{n} x_i = a\left(x_1 + x_2 + x_3 + ... + x_n\right) = ax_1 + ax_2 + ax_3 + ... + ax_n$$

$$b\sum_{i=1}^{n} x_i^2 = b\left(x_1^2 + x_2^2 + x_3^2 + ... + x_n^2\right) = bx_1^2 + bx_2^2 + bx_3^2 + ... + bx$$

Next we solve equations (B) and (D) simultaneously. The two equations are summarized below followed by their solution which was obtained using the method of elimination manually or using MATLAB with the command `solve` as illustrated in Chapter 10 on Solving Equations:

$$\sum_{i=1}^{n} y_i = na + b\sum_{i=1}^{n} x_i$$

$$\sum_{i=1}^{n} x_i y_i = a\sum_{i=1}^{n} x_i + b\sum_{i=1}^{n} x_i^2$$

The solution of the above two linear equations for the constants a and b is obtained as follows:

$$b = \frac{n\sum_{i=1}^{n} x_i y_i - \sum_{i=1}^{n} x_i \sum_{i=1}^{n} y_i}{n\sum_{i=1}^{n} x_i^2 - \left(\sum_{i=1}^{n} x_i\right)^2}$$

(I)

$$a = \frac{\sum_{i=1}^{n} x_i^2 \sum_{i=1}^{n} y_i - \sum_{i=1}^{n} x_i \sum_{i=1}^{n} x_i y_i}{n\sum_{i=1}^{n} x_i^2 - \left(\sum_{i=1}^{n} x_i\right)^2}$$

(II)

The procedure for curve fitting a straight line through the method of simple linear regression is summarized in the following five steps:

1. Calculate the sums $\sum_{i=1}^{n} x_i$, $\sum_{i=1}^{n} y_i$, $\sum_{i=1}^{n} x_i^2$, and $\sum_{i=1}^{n} x_i y_i$ using MATLAB (if calculated

manually, it is preferred to do it in a table). Table 14.1 shows each sum above along with the needed MATLAB command to compute it.

Table 14.1: MALTAB Command Needed to Compute Each Sum

Sum	MATLAB Command
x_i	`x`
y_i	`y`
$\sum_{i=1}^{n} x_i$	`sum(x)`
$\sum_{i=1}^{n} y_i$	`sum(y)`
$\sum_{i=1}^{n} x_i^2$	`mxx = x.*x` `sum(mxx)`
$\sum_{i=1}^{n} x_i y_i$	`mxy = x.*y` `sum(mxy)`
$\left(\sum_{i=1}^{n} x_i\right)^2$	`(sum(x))^2`

2. Substitute the above four sums into equations (I) and (II) to obtain the two constants a and b.

3. Using the two values of the constants a and b obtained above, write the equation of the best fitting line as follows:

$$y = a + bx$$

4. To check the results or to view the results graphically, plot the data using the MATLAB command `scatter`, then plot the best fitting line on the same graph using the MATLAB command `plot`. These two commands will be illustrated in the numerical example that will follow.

5. This last step is optional. We can calculate various types of errors between the data and the best fitting line, including the maximum error. However, this step will not be illustrated in the numerical example that will follow.

In order to illustrate simple linear regression in this chapter, we employ what is known as the Least Squares procedure. This procedure is used to obtain a best-fit curve or a best-fit line that represents the data based on simple linear regression. This is shown below using an example.

Let us now solve a numerical example using the above five steps. The following data (see Table 14.2) show the monthly salaries (x, in dollars) and the amount of monthly rent payments (y, in dollars) for a random sample of 16 employees at a certain company. Let us obtain the equation of the best fitting line using simple linear regression.

Table 14.2: Data for the Numerical Example

x	600	700	650	680	750	800	900	850
y	150	180	200	230	250	280	300	350

Table 14.2: Data for the Numerical Example (Continued)

x	950	1000	1020	1050	1100	1200	1250	1300
y	400	430	450	480	500	510	550	560

Let us now proceed to start the solution of the numerical example given above using the five steps outlined above for simple linear regression.

Step 1: Calculate the sums $\sum_{i=1}^{n} x_i$, $\sum_{i=1}^{n} y_i$, $\sum_{i=1}^{n} x_i^2$, and $\sum_{i=1}^{n} x_i y_i$ using the MATLAB command `sum` as follows:

1.1: Let us first input the data as follows:

```
>> x = [600 700 650 680 750 800 900
850 950 1000 1020 1050 1100 1200 1250
1300]

x =

  Columns 1 through 6

           600           700           650
680        750          800

  Columns 7 through 12

           900           850           950
1000       1020         1050

  Columns 13 through 16

          1100          1200          1250
1300
```

```
>> y = [150 180 200 230 250 280 300
350 400 430 450 480 500 510 550 560]

y =

  Columns 1 through 12

   150    180    200    230    250    280
300    350    400    430    450    480

  Columns 13 through 16

   500    510    550    560
```

1.2: Let us calculate the sum $\sum_{i=1}^{n} x_i$:

```
>> sumx = sum(x)

sumx =

       14800
```

1.3: Let us calculate the sum $\sum_{i=1}^{n} y_i$:

```
>> sumy = sum(y)

sumy =

        5820
```

1.4: Let us calculate the sum $\sum_{i=1}^{n} x_i^2$:

```
>> mxx = x.*x

mxx =

  Columns 1 through 5
```

```
       360000          490000          422500
462400        562500

  Columns 6 through 10

       640000          810000          722500
902500       1000000

  Columns 11 through 15

      1040400         1102500         1210000
1440000       1562500

  Column 16

      1690000

>> sumxx = sum(mxx)

sumxx =

     14417800
```

1.5: Let us calculate the sum $\sum_{i=1}^{n} x_i y_i$:

```
>> mxy = x.*y

mxy =

  Columns 1 through 5

        90000          126000          130000
156400        187500

  Columns 6 through 10

       224000          270000          297500
380000        430000

  Columns 11 through 15
```

```
          459000            504000            550000
612000          687500

    Column 16

          728000

>> sumxy = sum(mxy)

sumxy =

         5831900
```

1.6: Let us calculate the square sum $\left(\sum_{i=1}^{n} x_i\right)^2$:

```
>> sumx2 = sumx*sumx

sumx2 =

    219040000
```

1.7: Let us calculate the number of data points using the `size` command:

```
>> n = size(x)

n =

     1    16

>> n1 = n(2)

n1 =

    16
```

Step 2: Substitute the values of the above four sums into equations (I) and (II), evaluate the two expressions in these two equations, and obtain the values of the constants a and b. This is performed in MATLAB as follows:

```
>> a = (sumxx*sumy - sumx*sumxy)/(n1*sumxx - sumx2)

a =

 -206.1456

>> b = (n1*sumxy - sumx*sumy)/(n1*sumxx - sumx2)

b =

    0.6161
```

Step 3: Using the values of the constants a and b obtained above, we can finally write the equation of the best fitting line as follows:

$$y = -206.1456 + 0.6161x$$

Step 4: In order to compare the best fitting line with the data given, let us know plot the data along with the best fitting line on the same graph. This is performed using the two MATLAB commands `scatter` and `plot` as follows:

4.1: Let us calculate the y values using the best fit equation above as follows:

```
>> y1 = a + b*x

y1 =

  Columns 1 through 7

  163.5164   225.1268   194.3216
212.8047  255.9319  286.7371  348.3474

  Columns 8 through 14
```

```
    317.5423        379.1526        409.9577
422.2798  440.7629  471.5681  533.1784

    Columns 15 through 16

    563.9836  594.7887
```

4.2: Let us plot the best fitting straight line:

```
>> plot(x,y1)
>> hold on
```

4.3: Let us plot the given data points:

```
>> scatter(x,y)
```

4.4: Let us label the x-axis and the y-axis:

```
>> xlabel('x')
>> ylabel('y')
```

The graph generated by MATLAB is shown in Figure 14.1. It is clear from the figure that the generated best fitting line truly represents the data points given in the example.

Step 5: This error calculation step is optional and is not demonstrated here.

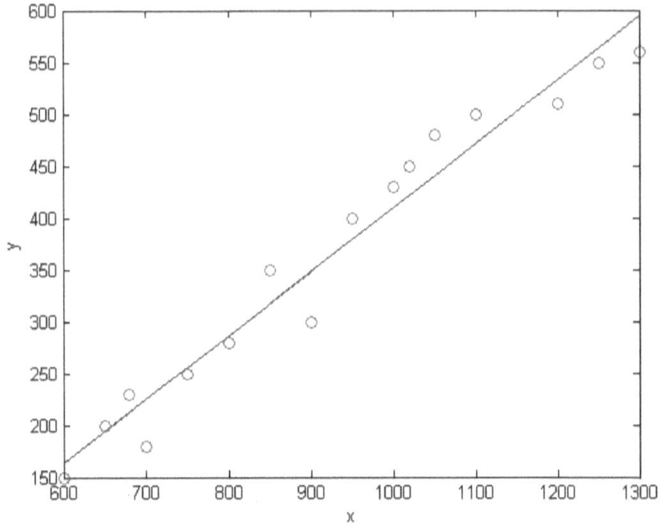

Figure 14.1: Best Fitting Line Along with the Given Data Points for the Example

Exercises

Solve the following three exercises on simple regression analysis. The first two should be done using simple linear regression using the five steps presented in this chapter. However, the third exercise is different. The third exercise is on the topic of simple nonlinear regression (specifically simple quadratic or simple parabolic regression) that is not illustrated in this chapter. However, enough information, hints, and steps are given in the exercise to make its solution attainable by the student.

1. A company makes a study of its assets (x, in thousands of dollars) and its

expenditure (y, in thousands of dollars) for a specific department. The following data are given (see Table 14.3):

Table 14.3: Data for Exercise 1

x	50	52	56	61	66	70	80	84	88	92
y	200	220	250	310	360	400	420	440	500	550

Find the equation of the best fitting for the above data. Plot the data along with the best fitting line on the same graph.

2. The following data (see Table 14.4) shows the expenditure x and the net income y, both in dollars, for a random sample of 12 firms.

Table 14.4: Data for Exercise 2

x	10	20	25	31	36	44	55	60	62	65	70	75
y	15	25	30	35	40	50	60	65	70	75	80	85

Find the equation of the best fitting for the above data. Plot the data along with the best fitting line on the same graph.

3. This exercise is optional for the beginning student and involves the equations of simple nonlinear regression, specifically what is called simple quadratic regression (also called simple parabolic regression). We will not solve a numerical problem in this exercise, but only want to write the

equations for this special case of simple nonlinear regression.

For a given set of data (x_i, y_i), assume the equation of the best fitting parabola to be quadratic in the general form:

$$y = a + bx + cx^2$$

Derive the three algebraic equations for the constant coefficients a, b, and c using the method of simple nonlinear regression.

Chapter 5

Regression Analysis – Part II

In this chapter we study multiple regression analysis. In this case, we have linear multiple regression and nonlinear multiple regression. We will derive the equations of linear multiple regression and illustrate how to use them using a numerical example[6].

Linear multiple regression is used to approximate the relationship between three or more variables. Two cases will be illustrated in this chapter: the first case involves three variables while the second case involves four variables. Both cases are based on the least squares technique of statistical analysis.

Consider a set of n data points (x_i, y_i, z_i) in three dimensions. In this case and since we are dealing with three-dimensional space, we obtain a best fitting plane instead of a best fitting line. The equation of the best fitting plane can be written in general as follows:

$$z = a + bx + cy$$

What is required is find the values of the three constants a, b, and c. The same procedure of Chapter 14 is followed here but note that we are dealing with three dimensions instead of two

[6] Multiple nonlinear regression is an advanced technique and will not be illustrated in this beginner's book.

dimensions. The above equation is repeated below with subscript $i = 1, 2, 3, \ldots\ldots n$:

$$z_i = a + bx_i + cy_i$$
(E)

It should be noted that the above equations is repeated n times, where n is the number of data points. Let us write the above equations for the first few data points as $i = 1, 2, 3, \ldots\ldots n$:

$$z_1 = a + bx_1 + cy_1$$

$$z_2 = a + bx_2 + cy_2$$

$$z_3 = a + bx_3 + cy_3$$

....

....

....

$$z_n = a + bx_n + cy_n$$

Let us now add the above list of equations to obtain the following expression:

$$z_1 + z_2 + z_3 + \ldots\ldots + z_n = a + bx_1 + cy_1 + a + bx_2 + cy_2 + a + bx_3 + cy_3 + \ldots\ldots + a + bx_n + cy_n$$

The above equation can be written in the following simplified form:

$$z_1 + z_2 + z_3 + \dots\dots + z_n = (a + a + a + \dots. + a)$$
$$+ (bx_1 + bx_2 + bx_3 + \dots\dots + bx_n) + (cy_1 + cy_2 + cy_3 + \dots\dots + cy_n)$$

where the sum of the a's in the first parenthesis is n times. The above equation can be written using the summation sign $\sum$ as follows:

$$\sum_{i=1}^{n} z_i = na + b\sum_{i=1}^{n} x_i + c\sum_{i=1}^{n} y_i$$

(F)

where the following summation notation is used in equation (F):

$$\sum_{i=1}^{n} z_i = z_1 + z_2 + z_3 + \dots\dots + z_n$$

$$na = a + a + a + \dots\dots + a \qquad (summed\ n\ times)$$

$$b\sum_{i=1}^{n} x_i = b(x_1 + x_2 + x_3 + \dots\dots + x_n) = bx_1 + bx_2 + bx_3 + \dots. + bx_n$$

$$c\sum_{i=1}^{n} y_i = c(y_1 + y_2 + y_3 + \dots\dots + y_n) = cy_1 + cy_2 + cy_3 + \dots. + cy_n$$

The next step would be to multiply equation (E) by x_i to obtain the following equation:

$$x_i z_i = a x_i + b x_i x_i + c x_i y_i$$
(G)

Writing the above equation out for each value $i = 1, 2, 3, \ldots\ldots n$, we obtain:

$$x_1 z_1 = a x_1 + b x_1^2 + c x_1 y_1$$

$$x_2 z_2 = a x_2 + b x_2^2 + c x_2 y_2$$

$$x_3 z_3 = a x_3 + b x_3^2 + c x_3 y_3$$

....

....

....

$$x_n z_n = a x_n + b x_n^2 + c x_n y_n$$

Adding the above list of equations results in the following expression:

$$x_1 z_1 + x_2 z_2 + x_3 z_3 + \ldots\ldots + x_n z_n = a x_1 + b x_1^2 + c x_1 y_1$$
$$+ a x_2 + b x_2^2 + c x_2 y_2 + a x_3 + b x_3^2 + c x_3 y_3 + \ldots\ldots + a x_n + b x_n^2 + c x_n y_n$$

The above equation can be simplified as follows:

$$x_1z_1 + x_2z_2 + x_3z_3 + ... + x_nz_n = ax_1 + ax_2 + ax_3 + ... + ax_n$$
$$+ bx_1^2 + bx_2^2 + bx_3^2 + ... + bx_n^2 + cx_1y_1 + cx_2y_2 + cx_3y_3 + ... + cx_ny_n$$

The above equation can be written using the summation sign $\sum$ as follows:

$$\sum_{i=1}^{n} x_iz_i = a\sum_{i=1}^{n} x_i + b\sum_{i=1}^{n} x_i^2 + c\sum_{i=1}^{n} x_iy_i$$
(H)

where the following summation notation is used in equation (H):

$$\sum_{i=1}^{n} x_iz_i = x_1z_1 + x_2z_2 + x_3z_3 + ... + x_nz_n$$

$$a\sum_{i=1}^{n} x_i = a(x_1 + x_2 + x_3 + ... + x_n) = ax_1 + ax_2 + ax_3 + ... + ax_n$$

$$b\sum_{i=1}^{n} x_i^2 = b(x_1^2 + x_2^2 + x_3^2 + ... + x_n^2) = bx_1^2 + bx_2^2 + bx_3^2 + ... + bx_n^2$$

$$c\sum_{i=1}^{n} x_iy_i = c(x_1y_1 + x_2y_2 + x_3y_3 + ... + x_ny_n)$$
$$= cx_1y_1 + cx_2y_2 + cx_3y_3 + ... + cx_ny_n$$

The next step would be to multiply equation (E) by y_i to obtain the following equation:

$$y_i z_i = a y_i + b x_i y_i + c y_i y_i$$
(J)

Writing the above equation out for each value $i = 1, 2, 3, \ldots\ldots n$, we obtain:

$$y_1 z_1 = a y_1 + b x_1 y_1 + c y_1^2$$

$$y_2 z_2 = a y_2 + b x_2 y_2 + c y_2^2$$

$$y_3 z_3 = a y_3 + b x_3 y_3 + c y_3^2$$

....

....

....

$$y_n z_n = a y_n + b x_n y_n + c y_n^2$$

Adding the above list of equations results in the following expression:

$$y_1 z_1 + y_2 z_2 + y_3 z_3 + \ldots\ldots + y_n z_n = a y_1 + b x_1 y_1 + c y_1^2 + a y_2 + b x_2 y_2 + c y_2^2 + a x_3 + b x_3 y_3 + c y_3^2 + \ldots\ldots + a y_n + b x_n y_n + cy$$

The above equation can be simplified as follows:

$$y_1 z_1 + y_2 z_2 + y_3 z_3 + \ldots\ldots + y_n z_n = a y_1 + a y_2 + a y_3 + \ldots\ldots + a y_n + b x_1 y_1 + b x_2 y_2 + b x_3 y_3 + \ldots\ldots + b x_n y_n + c y_1^2 + c y_2^2 + c y_3^2 + \ldots\ldots + cy$$

The above equation can be written using the summation sign $\sum$ as follows:

$$\sum_{i=1}^{n} y_i z_i = a\sum_{i=1}^{n} y_i + b\sum_{i=1}^{n} x_i y_i + c\sum_{i=1}^{n} y_i^2$$

(K)

where the following summation notation is used in equation (H):

$$\sum_{i=1}^{n} y_i z_i = y_1 z_1 + y_2 z_2 + y_3 z_3 + \ldots.. + y_n z_n$$

$$a\sum_{i=1}^{n} y_i = a(y_1 + y_2 + y_3 + \ldots\ldots + y_n) = ay_1 + ay_2 + ay_3 + \ldots\ldots + ay_n$$

$$b\sum_{i=1}^{n} x_i y_i = b(x_1 y_1 + x_2 y_2 + x_3 y_3 + \ldots\ldots + x_n y_n)$$

$$= bx_1 y_1 + bx_2 y_2 + bx_3 y_3 + \ldots\ldots + bx_n y_n$$

$$c\sum_{i=1}^{n} y_i^2 = c(y_1^2 + y_2^2 + y_3^2 + \ldots\ldots + y_n^2) = cy_1^2 + cy_2^2 + cy_3^2 + \ldots\ldots + cy_n^2$$

Next we list equations (F), (H) and (K) as shown in the boxed list below. These three equations should be solved simultaneously to obtain the three constants a, b, and c.

It is not advised to solve the three equations below symbolically and obtain a closed-form or analytical set of expressions for the three constants a, b, and c. Instead, it is advised to write out the three equations above explicitly for each specific set of data, then solve the three equations numerically using the solutions techniques illustrated in Chapter 10 on Solving Equations.

$$\sum_{i=1}^{n} z_i = na + b\sum_{i=1}^{n} x_i + c\sum_{i=1}^{n} y_i$$

(III)

$$\sum_{i=1}^{n} x_i z_i = a\sum_{i=1}^{n} x_i + b\sum_{i=1}^{n} x_i^2 + c\sum_{i=1}^{n} x_i y_i$$

(IV)

$$\sum_{i=1}^{n} y_i z_i = a\sum_{i=1}^{n} y_i + b\sum_{i=1}^{n} x_i y_i + c\sum_{i=1}^{n} y_i^2$$

(V)

The three equations above can be written out explicitly for each specific set of data points by calculating the various sums that appear in the equations. There are eight sums to be calculated in each case; namely $\sum_{i=1}^{n} x_i$, $\sum_{i=1}^{n} y_i$, $\sum_{i=1}^{n} z_i$, $\sum_{i=1}^{n} x_i y_i$, $\sum_{i=1}^{n} x_i z_i$, $\sum_{i=1}^{n} y_i z_i$, $\sum_{i=1}^{n} x_i^2$, and $\sum_{i=1}^{n} y_i^2$.

The procedure for curve fitting a plane through the method of multiple linear regression is summarized in the following four steps (one step from Chapter 14 is obviously missing; the step of showing the data and the best fit on a graph. It is not easy to show these data along with the best fit on a graph because we are now dealing with data in three-dimensional space):

1. Calculate the sums $\sum_{i=1}^{n} x_i$, $\sum_{i=1}^{n} y_i$, $\sum_{i=1}^{n} z_i$, $\sum_{i=1}^{n} x_i y_i$, $\sum_{i=1}^{n} x_i z_i$, $\sum_{i=1}^{n} y_i z_i$, $\sum_{i=1}^{n} x_i^2$, and $\sum_{i=1}^{n} y_i^2$ using MATLAB (if calculated manually, it is preferred to do it in a table). Table 15.1 shows each sum along with needed MATLAB command to compute it.

Table 15.1: MALATB Command Needed to Compute Each Sum

Sum	MATLAB Command
x_i	X
y_i	y
z_i	z
$\sum_{i=1}^{n} x_i$	`sum(x)`
$\sum_{i=1}^{n} y_i$	`sum(y)`
$\sum_{i=1}^{n} z_i$	`sum(z)`
$\sum_{i=1}^{n} x_i^2$	`mxx = x.*x` `sum(mxx)`
$\sum_{i=1}^{n} y_i^2$	`myy = y.*y` `sum(myy)`

$\sum_{i=1}^{n} x_i y_i$	`mxy = x.*y` `sum(mxy)`
$\sum_{i=1}^{n} x_i z_i$	`mxz = x.*z` `sum(mxz)`
$\sum_{i=1}^{n} y_i z_i$	`myz = y.*z` `sum(myz)`

2. Substitute the above eight sums into equations (III), (IV), and (V), and solve them, in order to obtain the three constants a, b, and c.

3. Using the three values of the constants a, b, and c obtained above, write the equation of the best fitting plane as follows:

   $$z = a + bx + cy$$

4. This last step is optional. We can calculate various types of errors between the data and the best fitting plane, including the maximum error. However, this step will not be illustrated in the numerical example that will follow.

Let us now solve a numerical example using the above four steps. The following data (see Table 15.2) is given in three dimensions. Let us obtain the equation of the best fitting plane using multiple linear regression.

Table 15.2: Data in Three Dimensions for the Numerical Example

x	120	150	210	290	350	400	500

y	10	12	17	25	30	33	38
z	105 0	200 0	210 0	150 0	210 0	330 0	110 0

The solution of this numerical example using MATLAB follows the four steps outlined above and is shown in detail below:

Step 1: Let us calculate the eight sums $\sum_{i=1}^{n} x_i$, $\sum_{i=1}^{n} y_i$, $\sum_{i=1}^{n} z_i$, $\sum_{i=1}^{n} x_i y_i$, $\sum_{i=1}^{n} x_i z_i$, $\sum_{i=1}^{n} y_i z_i$, $\sum_{i=1}^{n} x_i^2$, and $\sum_{i=1}^{n} y_i^2$ for the data given in Table 4 using MATLAB. The two basic MATLAB commands to be used are the command `sum` (summation command) and the command `.*` (term-by-term vector multiplication command). Both these commands were studied in great detail in Chapter 6 on vectors.

1.1: Let us enter the data points given in the Example:

```
>> x = [120 150 210 290 350 400 500]

x =

   120     150     210     290     350     400
500

>> y = [10 12 17 25 30 33 38]

y =
```

```
    10      12      17      25      30      33
38

>> z = [1050 2000 2100 1500 2100 3300
1100]

z =

  Columns 1 through 6

        1050          2000          2100
1500          2100          3300

  Column 7

        1100
```

1.2: Let us calculate the sum $\sum_{i=1}^{n} x_i$:

```
>> sumx = sum(x)

sumx =

        2020
```

1.3: Let us calculate the sum $\sum_{i=1}^{n} y_i$:

```
>> sumy = sum(y)

sumy =

   165
```

1.4: Let us calculate the sum $\sum_{i=1}^{n} z_i$:

```
>> sumz = sum(z)

sumz =
```

```
       13150
```

1.5: Let us calculate the sum $\sum_{i=1}^{n} x_i^2$:

```
>> mxx = x.*x

mxx =

  Columns 1 through 5

       14400         22500         44100
84100        122500

  Columns 6 through 7

      160000        250000

>> sumxx = sum(mxx)

sumxx =

      697600
```

1.6: Let us calculate the sum $\sum_{i=1}^{n} x_i y_i$:

```
>> mxy = x.*y

mxy =

  Columns 1 through 5

        1200          1800          3570
7250         10500

  Columns 6 through 7

       13200         19000
```

```
>> sumxy = sum(mxy)

sumxy =

       56520
```

1.7: Let us calculate the sum $\sum_{i=1}^{n} x_i z_i$:

```
>> mxz = x.*z

mxz =

  Columns 1 through 5

      126000         300000         441000
435000        735000

  Columns 6 through 7

     1320000        550000

>> sumxz = sum(mxz)

sumxz =

     3907000
```

1.8: Let us calculate the sum $\sum_{i=1}^{n} y_i z_i$:

```
>> myz = y.*z

myz =

  Columns 1 through 5

       10500          24000          35700
37500         63000

  Columns 6 through 7
```

```
        108900        41800

>> sumyz = sum(myz)

sumyz =

        321400
```

1.9: Let us calculate the sum $\sum_{i=1}^{n} y_i^2$:

```
>> myy = y.*y

myy =

  Columns 1 through 5

           100           144           289
625           900

  Columns 6 through 7

          1089          1444

>> sumyy = sum(myy)

sumyy =

          4591
```

1.10: Let us calculate the number of data points given in the Example using the `size` command:

```
>> n = size(x)

n =

     1     7

>> n1 = n(2)
```

```
n1 =

     7
```

Step 2: Write out the three equations (III), (IV), and (V) explicitly for the set of data given in Table 4 using the results of Step 1 above. Then solve the three resulting equations simultaneously for the constants a, b, and c using the MATLAB command `solve` that was illustrated in detail in Chapter 10 on Solving Equations.

The system of three simultaneous algebraic equations (III), (IV), and (V) can now be written for this example as follows:

$$7a + 2020b + 165c = 13150$$

$$2020a + 697600b + 56520c = 3907000$$

$$165a + 56520b + 4591c = 321400$$

Next, the above system of three simultanaeous algebraic equation is solve using the inverse matrix method in MATLAB as follows (check Chapter 10 on Solving Equations for details):

```
>> A = [n1 sumx sumy ; sumx sumxx
sumxy ; sumy sumxy sumyy]

A =

           7         2020          165
        2020       697600        56520
         165        56520         4591

>> b = [sumz ; sumxz ; sumyz]
```

```
b =

       13150
     3907000
      321400

>> inv(A)*b

ans =

   1.0e+03 *

    1.2850
   -0.0198
    0.2674

>> a = ans(1)

a =

   1.2850e+03

>> b = ans(2)

b =

  -19.7884

>> c = ans(3)

c =

  267.4393
```

Step 3: Using the values of the constants a, b, and c obtained from Step 2 above, write the equation of the best fitting plane as follows:

$$z = 1285 - 19.7884x + 267.4393y$$

Let us finally try to make a 3D plot of the best fitting plane along with the give data points. This is performed using MATLAB as follows:

```
>> z1 = a + b*x + c*y

z1 =

   1.0e+03 *

    1.5848          1.5260          1.6759
2.2324     2.3822     2.1951     1.5535

>> scatter(x,y,z)
>> hold on
>> plot3(x,y,z1)
>> xlabel('x')
>> ylabel('y')
>> zlabel('z')
```

Note that the MATLAB command `plot3` instead of the command `plot` is used above. The resulting graph is shown in Figure 15.1.

Step 4: This step of calculating errors is optional and is not illustrated for this example.

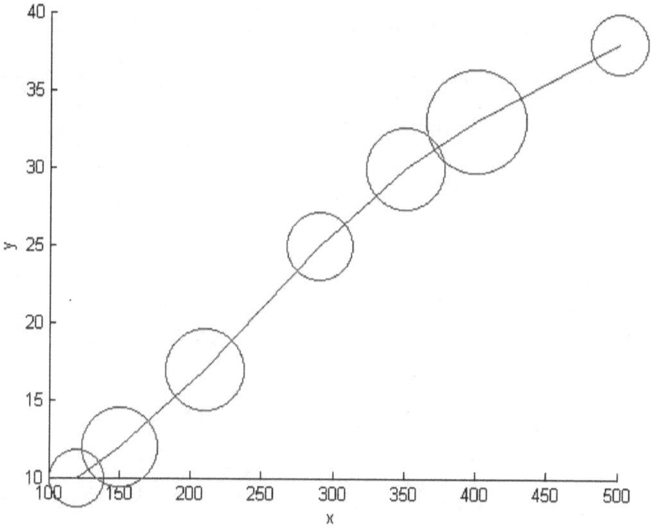

Figure 15.1: Best Fitting Plane and the Given Date Points for the Example

Next, we discuss what needs to be done if one wants to perform multiple linear regression on a set of data consisting of four variables in four-dimensional space. Given a set of n data points (x_i, y_i, z_i, w_i), one can follow the procedure of derivation given above to obtain the equation of the best fitting hyperplane in four-dimensional space. The equation of the best fitting hyperplane can be written in general as follows:

$$w = a + bx + cy + dz$$

We will not provide the details of the derivation, but list only the four simultaneous equations that need to be solved to obtain the four constants a, b, c, and d. The necessary four simultaneous equations are listed in the box below:

$$\sum_{i=1}^{n} w_i = na + b\sum_{i=1}^{n} x_i + c\sum_{i=1}^{n} y_i + d\sum_{i=1}^{n} z_i$$

(VI)

$$\sum_{i=1}^{n} x_i w_i = a\sum_{i=1}^{n} x_i + b\sum_{i=1}^{n} x_i^2 + c\sum_{i=1}^{n} x_i y_i + d\sum_{i=1}^{n} x_i z_i$$

(VII)

$$\sum_{i=1}^{n} y_i w_i = a\sum_{i=1}^{n} y_i + b\sum_{i=1}^{n} x_i y_i + c\sum_{i=1}^{n} y_i^2 + d\sum_{i=1}^{n} y_i z_i$$

(VIII)

$$\sum_{i=1}^{n} z_i w_i = a\sum_{i=1}^{n} z_i + b\sum_{i=1}^{n} x_i z_i + c\sum_{i=1}^{n} y_i z_i + d\sum_{i=1}^{n} z_i^2$$

(IX)

Next, if one considers the relationships between five variables in five-dimensional space, then one needs to write the equation of the best fitting hyperplane in five-dimensional space, then write a set of five simultaneous equations from which one can obtain the set of five constants needed for the equation of the hyperplane. This can be done by following the same procedure illustrated in this chapter for three variables.

Finally, the equations of multiple linear regression have been studied in detail in this chapter. On the other hand, the equations of multiple nonlinear regression are too complicated and are beyond the scope of this beginner's book. The interested reader is referred to the updated list of references at the end of this book for more details.

Exercises

Solve Exercises 1 and 2 following the four steps outlined in this chapter for multiple linear regression. Exercise 3 should be attempted only by advanced readers who are familiar with the derivation of the equations of multiple linear regression.

1. The following data (see Table 15.3) is give for three variables in three-dimensional space. Obtain the equation of the best fitting plane in three dimensions using multiple linear regression. Use MATLAB as illustrated in this chapter. (*Hint*: Use equations III, IV, and V).

Table 15.3: Data in Three Dimensions for Exercise 1

x	109	118	124	100	90	120	150
y	450	440	500	600	450	550	600
z	11	16	17	20	22	26	31

2. The following data (see Table 15.4) is give for four variables in four-dimensional space. Obtain the equation of the best fitting hyperplane in four dimensions using multiple linear regression. Use MATLAB as illustrated in this chapter. (*Hint*: Use equations VI, VII, VIII, and IX).

Table 15.4: Data in Three Dimensions for Exercise 2

x	234	245	257	270	250	270	299
y	2	4	7	9	7	11	13
z	890	810	910	900	1100	1200	1000
w	1030	1100	1200	1350	1440	1500	1570

3. Derive equations VI, VII, VIII, and XI in detail using multiple linear regression for four variables in four dimensions. This exercise should be attempted only by advanced readers (*Hint*: The detailed derivation can be found in the Solutions Manual at the end of this book).

Solutions to Exercises

1. Introduction

1. Perform the operation 3*4+6. The order of the operations will be discussed in subsequent chapters.

```
>> 3*4+6

ans =

    18
```

2. Perform the operation $\cos(5)$. The value of 5 is in radians.

```
>> cos(5)

ans =

    0.2837
```

3. Perform the operation $3\sqrt{6+x}$ for $x=4$.

```
>> x = 4

x =

     4

>> 3*sqrt(6+x)

ans =
```

```
        9.4868
```

4. Assign the value of 5.2 to the variable y.

```
>> y = 5.2

y =

        5.2000
```

5. Assign the values of 3 and 4 to the variables x and y, respectively, then calculate the value of z where $z = 2x - 7y$.

```
>> x = 3;
>> y = 4;
>> z = 2*x - 7*y

z =

      -22
```

6. Obtain help on the inv command.

```
>> help inv
 INV    Matrix inverse.
    INV(X) is the inverse of the
square matrix X.
    A warning message is printed
if X is badly scaled or
    nearly singular.

    See also slash, pinv, cond,
condest, lsqnonneg, lscov.

    Overloaded    functions    or
methods (ones with the same name
in other directories)
       help sym/inv.m
```

```
     Reference   page   in   Help
browser
        doc inv
```

7. Generate the following matrix: A

$$A = \begin{bmatrix} 1 & 0 & 2 & -3 \\ 0 & 5 & 2 & 2 \\ 1 & 2 & 3 & 4 \\ -2 & 0 & 1 & 3 \end{bmatrix}$$

```
>> A = [1 0 2 -3 ; 0 5 2 2 ; 1 2
3 4 ; -2 0 1 3]

A =

     1     0     2    -3
     0     5     2     2
     1     2     3     4
    -2     0     1     3
```

8. Generate the following vector b

$$b = \begin{Bmatrix} 1 \\ 2 \\ 3 \\ 4 \end{Bmatrix}$$

```
>> b = [1 ; 2 ; 3 ; 4]

b =

     1
     2
     3
```

```
    4
```

9. Evaluate the vector c where $\{c\} = [A]\{b\}$ where A is the matrix given in Exercise 7 above and b is the vector given in Exercise 8 above.

```
>> c = A*b

c =

    -5
    24
    30
    13
```

10. Solve the following system of simultaneous algebraic equation using Gaussian elimination.

$$\begin{bmatrix} 5 & 2 \\ 1 & 3 \end{bmatrix} \begin{Bmatrix} x_1 \\ x_2 \end{Bmatrix} = \begin{Bmatrix} 3 \\ -1 \end{Bmatrix}$$

```
>> A = [ 5 2 ; 1 3]

A =

     5     2
     1     3

>> b = [3 ; -1]

b =

     3
    -1

>> x = A\b

x =
```

```
    0.8462
   -0.6154
```

11. Solve the system of simultaneous algebraic equations of Exercise 10 above using matrix inversion.

```
>> x = inv(A)*b

x =

    0.8462
   -0.6154
```

12. Generate the following matrix X:

$$X = \begin{bmatrix} 1 & 0 & 6 \\ 1 & 2 & 3 \\ 4 & 5 & -2 \end{bmatrix}$$

```
>> X = [1 0 6 ; 1 2 3 ; 4 5 -2]

X =

     1     0     6
     1     2     3
     4     5    -2
```

13. Extract the sub-matrix in rows 2 to 3 and columns 1 to 2 of the matrix X in Exercise 12 above.

```
>> X(2:3,1:2)

ans =

     1     2
     4     5
```

14. Extract the second column of the matrix X in Exercise 12 above.

```
>> X(1:3,2)

ans =

     0
     2
     5
```

15. Extract the first row of the matrix X in Exercise 12 above.

```
>> X(1,1:3)

ans =

     1     0     6
```

16. Extract the element in row 1 and column 3 of the matrix X in Exercise 12 above.

```
>> X(1,3)

ans =

     6
```

17. Generate the row vector x with integer values ranging from 1 to 9.

```
>> x = [1 2 3 4 5 6 7 8 9]

x =

     1     2     3     4     5
6     7     8     9
```

18. Plot the graph of the function $y = x^3 - 2$ for the range of the values of x in Exercise 17 above.

```
>> y = x.^3 - 2

y =

     -1       6      25      62     123
214    341    510    727

>> plot(x,y)
```

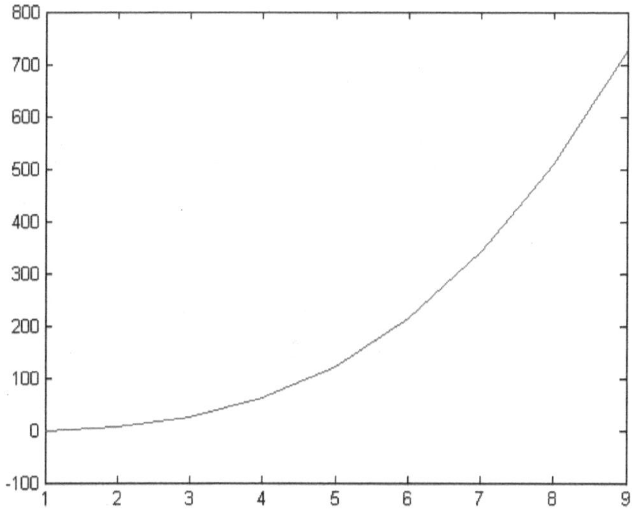

19. Generate a 4 x 4 magic square. What is the total of each row, column, and diagonal in this matrix.

```
>> magic(4)

ans =

     16      2      3     13
```

```
5     11     10      8
9      7      6     12
4     14     15      1
```

The sum of each row, column, or diagonal is 34.

2. Statistics Review – Part I

```
    1. >> x = [25 18 17 24 11 10 20
          30 25 12 11 12 15 28 33 20 12
          20 22 30 15 18 22 19 31]

x =

  Columns 1 through 11

    25       18       17       24       11
10     20     30     25     12     11

  Columns 12 through 22

    12       15       28       33       20
12     20     22     30     15     18

  Columns 23 through 25

    22     19     31


    2. >> size(x)

ans =

     1     25


    3. >> mean(x)

ans =
```

```
    20

>> median(x)

ans =

    20

>> mode(x)

ans =

    12
```

9. ```
>> range(x)

ans =

    23

>> iqr(x)

ans =

   10.7500

>> var(x)

ans =

   47.9167

>> std(x)

ans =

    6.9222
```

9. ```
>> max(x)
```

```
ans =

   33

>> min(x)

ans =

   10
```

9. `>> quantile(x,0.30)`

```
ans =

    15

>> quantile(x,0.50)

ans =

    20

>> quantile(x,0.70)

ans =

    24
```

9. `>> prctile(x,47)`

```
ans =

   19.2500

>> prctile(x,55)

ans =

    20

>> prctile(x,80)
```

```
ans =

   26.5000


9.          >> zscore(x)

ans =

  Columns 1 through 7

    0.7223     -0.2889     -0.4334
0.5779      -1.3002      -1.4446
0

  Columns 8 through 14

    1.4446      0.7223     -1.1557
-1.3002      -1.1557      -0.7223
1.1557

  Columns 15 through 21

    1.8780           0     -1.1557
0    0.2889    1.4446   -0.7223

  Columns 22 through 25

   -0.2889      0.2889     -0.1445
1.5891


9.          >>  x  =  [1.24  2.56
            3.45  4.66  3.56  4.90
            3.33  2.45  2.34  2.00
            1.45 1.56]

x =

  Columns 1 through 7
```

```
    1.2400       2.5600       3.4500
4.6600       3.5600       4.9000
3.3300

  Columns 8 through 12

    2.4500       2.3400       2.0000
1.4500    1.5600

>> y = [600 450 320 300 430 330
470 570 340 200 250 390]

y =

  Columns 1 through 11

   600     450     320     300     430
330     470     570     340     200
250

  Column 12

   390

>> corrcoef(x,y)

ans =

    1.0000    -0.1906
   -0.1906     1.0000
```

Thus the correlation coefficient in this Exercise turns out to be -0.1906 indicating a weak negative correlation. In the next Exercise, the same correlation coefficient is calculated from scratch, but this time using the formula of Pearson's correlation coefficient (see Chapter 12).

10. <u>Step One:</u> Enter the data points given in the Exercise:

```
>> x = [1.24 2.56 3.45 4.66 3.56
4.90 3.33 2.45 2.34 2.00 1.45 1.56]
```

```
x =

  Columns 1 through 7

    1.2400        2.5600        3.4500
4.6600        3.5600        4.9000
3.3300

  Columns 8 through 12

    2.4500        2.3400        2.0000
1.4500    1.5600

>> y = [600 450 320 300 430 330 470
570 340 200 250 390]

y =

  Columns 1 through 11

   600     450     320     300     430
330    470    570    340    200    250

  Column 12

   390
```

Step Two: Calculate the five different sums needed in the formula for the correlation coefficient:

```
>> sumx = sum(x)

sumx =

   33.5000

>> sumy = sum(y)

sumy =

        4650
```

```
>> mxx = x.*x

mxx =

  Columns 1 through 7

    1.5376        6.5536       11.9025
21.7156        12.6736        24.0100
11.0889

  Columns 8 through 12

    6.0025        5.4756        4.0000
2.1025    2.4336

>> sumxx = sum(mxx)

sumxx =

  109.4960

>> myy = y.*y

myy =

  Columns 1 through 5

      360000                      202500
102400        90000       184900

  Columns 6 through 10

      108900                      220900
324900      115600        40000

  Columns 11 through 12

       62500       152100

>> sumyy = sum(myy)

sumyy =
```

```
     1964700

>> mxy = x.*y

mxy =

   1.0e+03 *

  Columns 1 through 7

    0.7440        1.1520        1.1040
1.3980          1.5308          1.6170
1.5651

  Columns 8 through 12

    1.3965        0.7956        0.4000
0.3625    0.6084

>> sumxy = sum(mxy)

sumxy =

   1.2674e+04

>> sumx2 = sumx*sumx

sumx2 =

   1.1223e+03

>> sumy2 = sumy*sumy

sumy2 =

    21622500

>> n = size(x)

n =

     1    12
```

```
>> n1 = n(2)

n1 =

    12
```

Step Three: Use the formula to compute the value of the correlation coefficient based on the sums obtained above:

```
>> r = (n1*sumxy - sumx*sumy) / (sqrt(n1*sumxx-
sumx2)*sqrt(n1*sumyy-sumy2))

r =

   -0.1906
```

It is clear from the above result that the correlation between the two data sets is negative and weak. The value of the correlation coefficient is close to 0 indicating a weak correlation. It is the same value obtained in Exercise 9 above that was obtained directly using the MATLAB command `corrcoef`.

11. The same value of the correlation coefficient is obtained in both Exercises 9 and 10 above. The first value in Exercise 9 was obtained directly using the MATLAB command `corrcoef`. The second value in Exercise 10 was obtained from scratch using the formula for Pearson's correlation coefficient.

3. Statistics Review – Part II

1.
```
>> x = [25 18 17 24 11 10 20 30 25
12 11 12 15 28 33 20 12 20 22 30 15
18 22 19 31]
```

```
x =

  Columns 1 through 11

    25      18      17      24      11      10
20    30    25    12    11

  Columns 12 through 22

    12      15      28      33      20      12
20    22    30    15    18

  Columns 23 through 25

    22    19    31

>> bar(x)
>> hold on
>> xlabel('Data')
>> ylabel('Number')
```

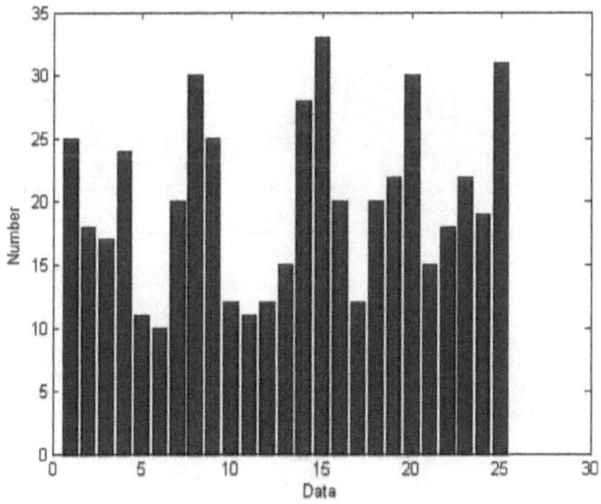


```
   2. >> hist(x)
>> hold on
>> xlabel('Data')
>> ylabel('Number')
```

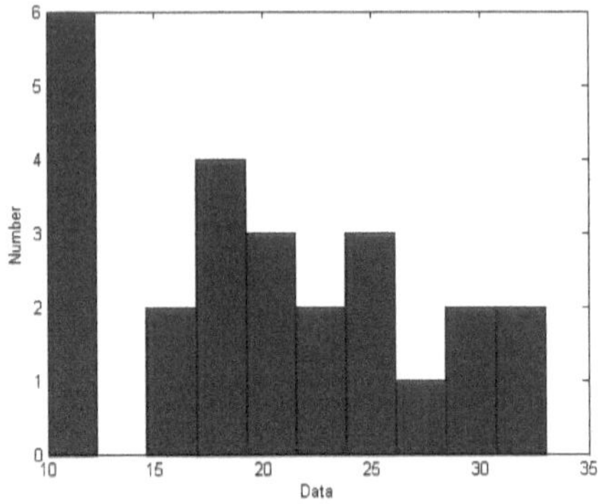

3. >> cdfplot(x)

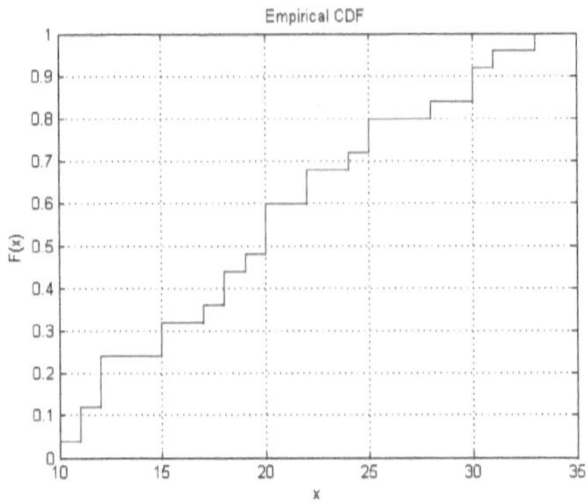

4. >> size(x)

```
ans =

     1    25

>> y = 1:25

y =

  Columns 1 through 11
```

```
     1         2         3         4         5         6
7        8        9       10       11

  Columns 12 through 22

    12        13        14        15        16        17
18       19        20       21       22

  Columns 23 through 25

    23       24       25

>> scatter(y,x)
>> hold on
>> xlabel('Data')
>> ylabel('Number')
```

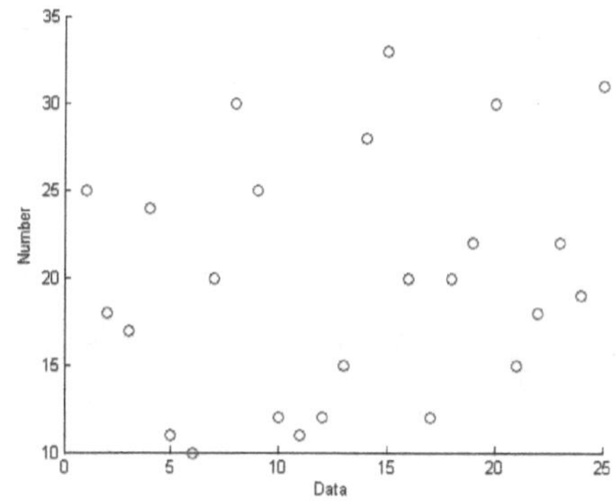

5. >> pie(x)

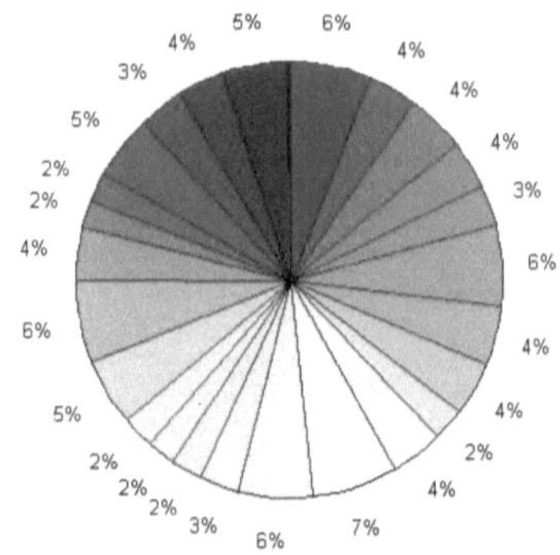

6. >> rose(x)

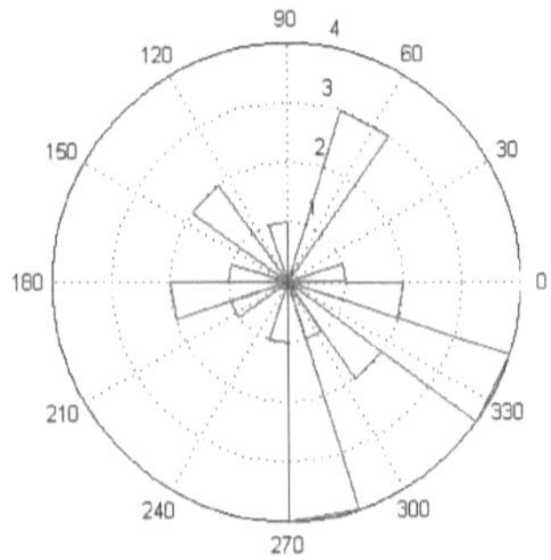

7. >> stem(x)

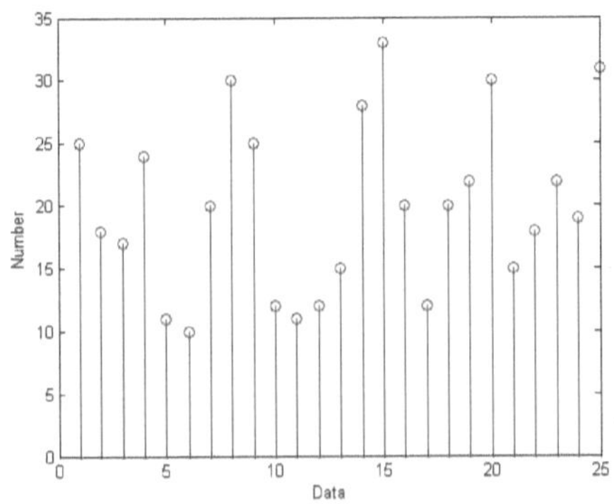

```
8. >> zscore(x)

ans =

  Columns 1 through 7

    0.7223   -0.2889   -0.4334    0.5779
-1.3002   -1.4446         0

  Columns 8 through 14

    1.4446    0.7223   -1.1557   -1.3002
-1.1557   -0.7223    1.1557

  Columns 15 through 21

    1.8780         0   -1.1557         0
0.2889    1.4446   -0.7223

  Columns 22 through 25

   -0.2889    0.2889   -0.1445    1.5891

>> scatter(y,ans)
>> hold on
>> xlabel('Data')
>> ylabel('zscores')
```

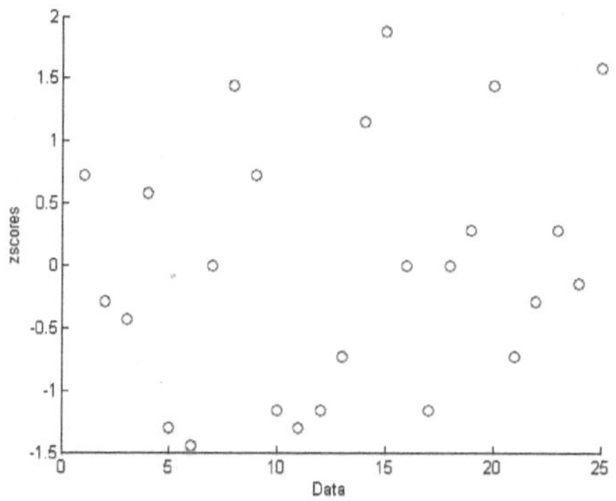

```
   9. >> boxplot(x)
>> hold on
>> ylabel('Data')
```

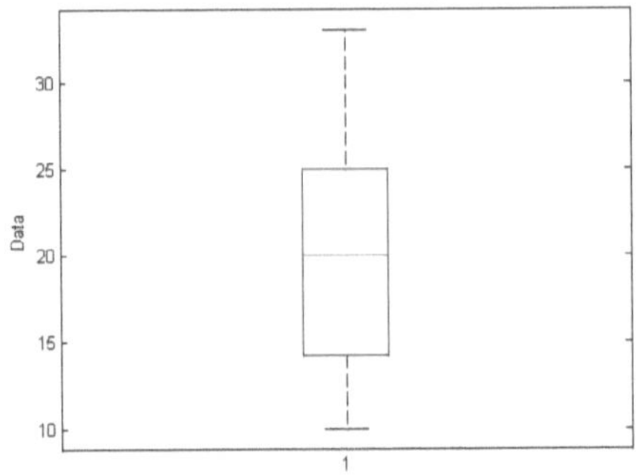

4. Regression Analysis - Part I

1. Step One: Enter the given data points and calculate the different required sums:

```
>> x = [50 52 56 61 66 70 80 84 88 92]

x =

    50      52      56      61      66      70
80      84      88      92

>> y = [200 220 250 310 360 400 420 440
500 550]

y =

   200     220     250     310     360     400
420     440     500     550

>> sumx = sum(x)

sumx =
```

```
   699

>> sumy = sum(y)

sumy =

        3650

>> mxx = x.*x

mxx =

  Columns 1 through 6

        2500             2704             3136
3721        4356        4900

  Columns 7 through 10

        6400             7056             7744
8464

>> sumxx = sum(mxx)

sumxx =

       50981

>> mxy = x.*y

mxy =

  Columns 1 through 5

       10000            11440            14000
18910          23760

  Columns 6 through 10

       28000            33600            36960
44000          50600
```

```
>> sumxy = sum(mxy)

sumxy =

      271270

>> sumx2 = sumx*sumx

sumx2 =

      488601

>> n = size(x)

n =

     1    10

>> n1 = n(2)

n1 =

    10
```

Step Two: Calculate the two constants *a* and *b* using the two formulas given:

```
>> a = (sumxx*sumy - sumx*sumxy)/(n1*sumxx - sumx2)

a =

 -166.7726

>> b = (n1*sumxy - sumx*sumy)/(n1*sumxx - sumx2)

b =

    7.6076
```

Step Three: Write the equation of the best fitting line as follows:

$$y = -166.7726 + 7.6076x$$

Step Four: Plot the best fitting line along with the give data points on the same graph for comparison:

```
>> y1 = a + b*x

y1 =

  Columns 1 through 7

  213.6084   228.8236   259.2541   297.2922
335.3303  365.7608  441.8370

  Columns 8 through 10

  472.2674  502.6979  533.1284

>> plot(x,y1)
>> hold on
>> scatter(x,y)
>> xlabel('x')
>> ylabel('y')
```

The resulting graph is shown in Figure 14.2.

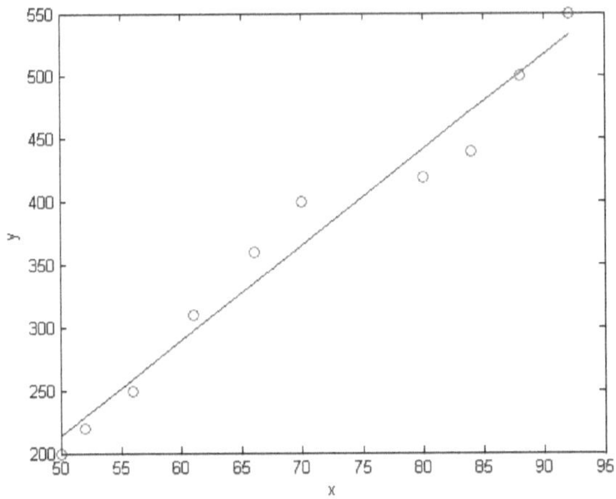

2. Step One: Enter the given data points and calculate the different sums needed:

```
>> x = [10 20 25 31 36 44 55 60 62 65 70
75]

x =

    10      20      25      31      36      44
55      60      62      65      70      75

>> y = [15 25 30 35 40 50 60 65 70 75 80
85]

y =

    15      25      30      35      40      50
60      65      70      75      80      85

>> sumx = sum(x)

sumx =

   553
```

```
>> sumy = sum(y)

sumy =

   630

>> mxx = x.*x

mxx =

  Columns 1 through 6

        100              400              625
961        1296         1936

  Columns 7 through 12

       3025             3600             3844
4225         4900         5625

>> sumxx = sum(mxx)

sumxx =

       30537

>> mxy = x.*y

mxy =

  Columns 1 through 5

        150              500              750
1085         1440

  Columns 6 through 10

       2200             3300             3900
4340         4875

  Columns 11 through 12

       5600         6375
```

```
>> sumxy = sum(mxy)

sumxy =

       34515

>> sumx2 = sumx*sumx

sumx2 =

      305809

>> n = size(x)

n =

     1    12

>> n1 = n(2)

n1 =

    12
```

Step Two: Calculate the two constants *a* and *b* as follows:

```
>>       a       =       (sumxx*sumy       -
sumx*sumxy)/(n1*sumxx - sumx2)

a =

    2.4988

>> b = (n1*sumxy - sumx*sumy)/(n1*sumxx -
sumx2)

b =

    1.0850
```

Step 3: Write the equation of the best fitting line as follows:

$$y = 2.4988 + 1.0850x$$

Step 4: Plot the best fitting line along with the given date points on the same graph for comparison:

```
>> y1 = a + b*x

y1 =

  Columns 1 through 7

   13.3490    24.1991    29.6242    36.1343
41.5594   50.2395   62.1747

  Columns 8 through 12

   67.5998    69.7699    73.0249    78.4500
83.8751

>> plot(x,y1)
>> hold on
>> scatter(x,y)
>> xlabel('x')
>> ylabel('y')
```

The resulting graph is shown in Figure 14.3.

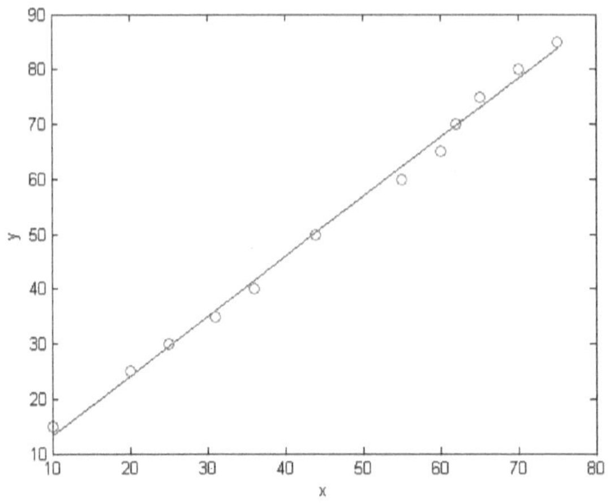

3.. Let us start with the quadratic equation given in the Exercise as follows:

$$y = a + bx + cx^2$$

where a, b, and c are constant coefficients to be determined for a specific set of data. We will proceed with the detailed derivation of the equations of simple nonlinear regression.

For a set of data x_i and y_i (where $i = 1,2,3,......n$), the above equation becomes for each point (x_i, y_i):

$$y_i = a + bx_i + cx_i^2$$
(AA)

It should be noted that the above equations is repeated n times, where n is the number of data points. Let us write the above equations for the first few data points as $i = 1,2,3,......n$:

$$y_1 = a + bx_1 + cx_1^2$$

$$y_2 = a + bx_2 + cx_2^2$$

$$y_3 = a + bx_3 + cx_3^2$$

....

....

....

$$y_n = a + bx_n + cx_n^2$$

Let us now add the above list of equations to obtain the following expression:

$$y_1 + y_2 + y_3 + \dots\dots + y_n = a + bx_1 + cx_1^2 + a + bx_2 + cx_2^2 + a + bx_3 + cx_3^2 + \dots\dots + a + bx_n + cx_n^2$$

The above equation can be written in the following simplified form:

$$y_1 + y_2 + y_3 + \dots\dots + y_n = (a + a + a + \dots\dots + a) + (bx_1 + bx_2 + bx_3 + \dots\dots + bx_n) + (cx_1^2 + cx_2^2 + cx_3^2 + \dots\dots + cx_n^2)$$

where the sum of the a's in the first parenthesis is n times. The above equation can be written using the summation sign $\sum$ as follows:

$$\sum_{i=1}^{n} y_i = na + b\sum_{i=1}^{n} x_i + c\sum_{i=1}^{n} x_i^2$$
(BB)

where the following summation notation is used in equation (BB):

$$\sum_{i=1}^{n} y_i = y_1 + y_2 + y_3 + \ldots\ldots.. + y_n$$

$$na = a + a + a + \ldots.. + a \qquad (summed\ \ n\ \ times)$$

$$b\sum_{i=1}^{n} x_i = b(x_1 + x_2 + x_3 + \ldots\ldots + x_n) = bx_1 + bx_2 + bx_3 + \ldots. + bx_n$$

$$c\sum_{i=1}^{n} x_i^2 = c(x_1^2 + x_2^2 + x_3^2 + \ldots.. + x_n^2) = cx_1^2 + cx_2^2 + cx_3^2 + \ldots.. + cx_n^2$$

The next step would be to multiply equation (AA) by x_i to obtain the following equation:

$$x_i y_i = ax_i + bx_i x_i + cx_i^2 x_i = ax_i + bx_i^2 + cx_i^3$$
(CC)

Writing the above equation out for each value $i = 1, 2, 3, \ldots\ldots n$, we obtain:

$$x_1 y_1 = ax_1 + bx_1^2 + cx_1^3$$

$$x_2 y_2 = ax_2 + bx_2^2 + cx_2^3$$

$$x_3 y_3 = a x_3 + b x_3^2 + c x_3^3$$

....

....

....

$$x_n y_n = a x_n + b x_n^2 + c x_n^3$$

Adding the above list of equations results in the following expression:

$$x_1 y_1 + x_2 y_2 + x_3 y_3 + \ldots\ldots + x_n y_n = a x_1 + b x_1^2 + c x_1^3 + a x_2 + b x_2^2 + c x_2^3 + a x_3 + b x_3^2 + c x_3^3 + \ldots\ldots\ldots + a x_n + b x_n^2 + c x_n^3$$

The above equation can be simplified as follows:

$$x_1 y_1 + x_2 y_2 + x_3 y_3 + \ldots\ldots + x_n y_n = a x_1 + a x_2 + a x_3 + \ldots\ldots + a x_n + b x_1^2 + b x_2^2 + b x_3^2 + \ldots\ldots + b x_n^2 + c x_1^3 + c x_2^3 + c x_3^3 + \ldots\ldots + c x_n^3$$

The above equation can be written using the summation sign $\sum$ as follows:

$$\sum_{i=1}^{n} x_i y_i = a \sum_{i=1}^{n} x_i + b \sum_{i=1}^{n} x_i^2 + c \sum_{i=1}^{n} x_i^3$$

(DD)

where the following summation notation is used in equation (DD):

$$\sum_{i=1}^{n} x_i y_i = x_1 y_1 + x_2 y_2 + x_3 y_3 + \ldots.. + x_n y_n$$

$$a\sum_{i=1}^{n} x_i = a\left(x_1 + x_2 + x_3 + \ldots\ldots + x_n\right) = ax_1 + ax_2 + ax_3 + \ldots\ldots + ax_n$$

$$b\sum_{i=1}^{n} x_i^2 = b\left(x_1^2 + x_2^2 + x_3^2 + \ldots\ldots + x_n^2\right) = bx_1^2 + bx_2^2 + bx_3^2 + \ldots\ldots + bx$$

$$c\sum_{i=1}^{n} x_i^3 = c\left(x_1^3 + x_2^3 + x_3^3 + \ldots\ldots + x_n^3\right) = cx_1^3 + cx_2^3 + cx_3^3 + \ldots\ldots + cx_n^3$$

The next step would be to multiply equation (AA) by x_i^2 to obtain the following equation:

$$x_i^2 y_i = ax_i^2 + bx_i x_i^2 + cx_i^2 x_i^2 = ax_i^2 + bx_i^3 + cx_i^4$$
(EE)

Writing the above equation out for each value $i = 1, 2, 3, \ldots\ldots n$, we obtain:

$$x_1^2 y_1 = ax_1^2 + bx_1^3 + cx_1^4$$

$$x_2^2 y_2 = ax_2^2 + bx_2^3 + cx_2^4$$

$$x_3^2 y_3 = ax_3^2 + bx_3^3 + cx_3^4$$

....

....

....

$$x_n^2 y_n = ax_n^2 + bx_n^3 + cx_n^4$$

Adding the above list of equations results in the following expression:

$$x_1^2 y_1 + x_2^2 y_2 + x_3^2 y_3 + ... + x_n^2 y_n = ax_1^2 + bx_1^3 + cx_1^4 + ax_2^2 + bx_2^3 + cx_2^4 \\ + ax_3^2 + bx_3^3 + cx_3^4 + ... + ax_n^2 + bx_n^3 + cx_n^4$$

The above equation can be simplified as follows:

$$x_1^2 y_1 + x_2^2 y_2 + x_3^2 y_3 + ... + x_n^2 y_n = ax_1^2 + ax_2^2 + ax_3^2 + ... + ax_n^2 \\ + bx_1^3 + bx_2^3 + bx_3^3 + ... + bx_n^3 + cx_1^4 + cx_2^4 + cx_3^4 + ... + cx_n^4$$

The above equation can be written using the summation sign $\sum$ as follows:

$$\sum_{i=1}^{n} x_i^2 y_i = a\sum_{i=1}^{n} x_i^2 + b\sum_{i=1}^{n} x_i^3 + c\sum_{i=1}^{n} x_i^4$$

(FF)

where the following summation notation is used in equation (FF):

$$\sum_{i=1}^{n} x_i^2 y_i = x_1^2 y_1 + x_2^2 y_2 + x_3^2 y_3 + ... + x_n^2 y_n$$

$$a\sum_{i=1}^{n} x_i^2 = a\left(x_1^2 + x_2^2 + x_3^2 + ... + x_n^2\right) = ax_1^2 + ax_2^2 + ax_3^2 + ... + ax_n^2$$

$$b\sum_{i=1}^{n} x_i^3 = b\left(x_1^3 + x_2^3 + x_3^3 + ... + x_n^3\right) = bx_1^3 + bx_2^3 + bx_3^3 + ... + bx_n^3$$

$$c\sum_{i=1}^{n} x_i^4 = c\left(x_1^4 + x_2^4 + x_3^4 + ... + x_n^4\right) = cx_1^4 + cx_2^4 + cx_3^4 + ... + cx_n^4$$

Next we solve equations (BB), (DD), and (FF) simultaneously. The three equations are summarized below. The three equations can then be solve simultaneously for the three constants a, b, and c. However, this is not performed here as it is not required by the Exercise. Usually, the various sums are computed first, the inserted into the three equations directly. This is followed by solving the three equations numerically using the MATLAB command `solve` as explained in Chapter 10 on Solving Equations.

$$\sum_{i=1}^{n} y_i = na + b\sum_{i=1}^{n} x_i + c\sum_{i=1}^{n} x_i^2$$

$$\sum_{i=1}^{n} x_i y_i = a\sum_{i=1}^{n} x_i + b\sum_{i=1}^{n} x_i^2 + c\sum_{i=1}^{n} x_i^3$$

$$\sum_{i=1}^{n} x_i^2 y_i = a\sum_{i=1}^{n} x_i^2 + b\sum_{i=1}^{n} x_i^3 + c\sum_{i=1}^{n} x_i^4$$

5. Regression Analysis - Part II

1. Step One: Enter the given data points and calculate the eight different sums needed as follows:

   1.1: Enter the given data points:

```
>> x = [109 118 124 100 90 120 150]

x =

   109     118     124     100      90
120   150

>> y = [450 440 500 600 450 550
600]

y =

   450     440     500     600     450
550   600

>> z = [11 16 17 20 22 26 31]

z =

    11      16      17      20      22
26    31
```

   1.2: Calculate the eight different sums needed for this exercise:

```
>> sumx = sum(x)

sumx =

   811

>> sumy = sum(y)
```

```
sumy =

        3590

>> sumz = sum(z)

sumz =

   143

>> mxx = x.*x

mxx =

  Columns 1 through 5

       11881                       13924
15376        10000         8100

  Columns 6 through 7

       14400        22500

>> sumxx = sum(mxx)

sumxx =

       96181

>> mxy = x.*y

mxy =

  Columns 1 through 5

       49050                       51920
62000        60000        40500

  Columns 6 through 7

       66000        90000

>> sumxy = sum(mxy)
```

```
sumxy =

      419470

>> mxz = x.*z

mxz =

  Columns 1 through 5

        1199                                    1888
2108          2000          1980

  Columns 6 through 7

        3120          4650

>> sumxz = sum(mxz)

sumxz =

       16945

>> myz = y.*z

myz =

  Columns 1 through 5

        4950                                    7040
8500        12000          9900

  Columns 6 through 7

       14300         18600

>> sumyz = sum(myz)

sumyz =

       75290
```

```
>> myy = y.*y

myy =

  Columns 1 through 5

      202500                                     193600
250000        360000        202500

  Columns 6 through 7

      302500        360000

>> sumyy = sum(myy)

sumyy =

     1871100

>> n = size(x)

n =

     1     7

>> n1 = n(2)

n1 =

     7
```

Step Two: Write the system of three simultaneous governing equations for the three constants a, b, and c as follows:

$$7a + 811b + 3590c = 143$$

$$811a + 96181b + 419470c = 16945$$

$$3590a + 419470b + 1891100c = 75290$$

Step Three: Solve the system of three simultaneous algebraic equations and obtain the values of the three constants a, b, and c as follows:

```
>> A = [n1 sumx sumy ; sumx sumxx
sumxy ; sumy sumxy sumyy]

A =

            7                        811
3590
          811                      96181
419470
         3590                     419470
1871100

>> b = [sumz ; sumxz ; sumyz]

b =

          143
        16945
        75290

>> inv(A)*b

ans =

   -17.4828
     0.0813
     0.0555

>> a = ans(1)

a =

   -17.4828

>> b = ans(2)

b =
```

```
        0.0813

>> c = ans(3)

c =

        0.0555
```

Step Four: Write the equation of the best fitting plane as follows:

$$z = -17.4828 + 0.0813x + 0.0555y$$

No need to plot a graph of the plane or estimate the error in the analysis. But we can calculate the values of z using the best fitting plane equation as follows:

```
>> z1 = a + b*x + c*y

z1 =

   16.3792        16.5558        20.3767
23.9795        14.8339         22.8287
28.0462
```

The approximate calculated values shown above are reasonable and are very close to the data values given in the exercise.

2. <u>Step One:</u> Let us enter the data points given as well as calculate the twelve different sums needed:

   2.1: Let us first enter the data points given in the Exercise:

```
>> x = [234 245 257 270 250 270
299]

x =
```

```
    234      245      257      270      250
270   299

>> y = [2 4 7 9 7 11 13]

y =

     2        4        7        9        7
11     13

>> z = [890 810 910 900 1100 1200
1000]

z =

  Columns 1 through 6

         890                          810
910              900                  1100
1200

  Column 7

        1000

>> w = [1030 1100 1200 1350 1440
1500 1570]

w =

  Columns 1 through 6

        1030                         1100
1200             1350                1440
1500

  Column 7

        1570
```

2.2: Let us calculate the twelve different sums needed in this Exercise:

```
>> sumx = sum(x)

sumx =

        1825

>> sumy = sum(y)

sumy =

    53

>> sumz = sum(z)

sumz =

        6810

>> sumw = sum(w)

sumw =

        9190

>> mxx = x.*x

mxx =

  Columns 1 through 5

       54756                          60025
66049        72900        62500

  Columns 6 through 7

       72900        89401

>> sumxx = sum(mxx)

sumxx =

      478531
```

```
>> myy = y.*y

myy =

     4      16      49      81      49
121   169

>> sumyy = sum(myy)

sumyy =

   489

>> mzz = z.^z

mzz =

   Inf    Inf    Inf    Inf    Inf
Inf   Inf

>> mzz = z.*z

mzz =

  Columns 1 through 5

      792100                    656100
828100       810000     1210000

  Columns 6 through 7

     1440000     1000000

>> sumzz = sum(mzz)

sumzz =

     6736300

>> mxy = x.*y

mxy =
```

```
  Columns 1 through 5

         468                                  980
  1799          2430           1750

  Columns 6 through 7

        2970          3887

>> sumxy = sum(mxy)

sumxy =

       14284

>> mxz = x.*z

mxz =

  Columns 1 through 5

      208260                             198450
  233870       243000       275000

  Columns 6 through 7

      324000       299000

>> sumxz = sum(mxz)

sumxz =

     1781580

>> myz = y.*z

myz =

  Columns 1 through 5

        1780                                 3240
  6370          8100           7700
```

```
  Columns 6 through 7

       13200        13000

>> sumyz = sum(myz)

sumyz =

       53390

>> mxw = x.*w

mxw =

  Columns 1 through 5

      241020                              269500
308400        364500        360000

  Columns 6 through 7

      405000        469430

>> sumxw = sum(mxw)

sumxw =

     2417850

>> myw = y.*w

myw =

  Columns 1 through 5

        2060                              4400
8400        12150        10080

  Columns 6 through 7

       16500        20410

>> sumyw = sum(myw)
```

```
sumyw =

        74000

>> mzw = z.*w

mzw =

  Columns 1 through 5

      916700                              891000
1092000      1215000      1584000

  Columns 6 through 7

     1800000      1570000

>> sumzw = sum(mzw)

sumzw =

     9068700

>> n =size(x)

n =

     1     7

>> n1 = n(2)

n1 =

     7
```

Step Two: Let us write the system of four simultaneous algebraic equations for the four constants a, b, c, and d as follows:

$$7a + 1825b + 53c + 6810d = 9190$$

$$1825a + 478531b + 14284c + 1781580d = 2417850$$

$$53a + 14284b + 489c + 53390d = 74000$$

$$6810a + 1781580b + 53390c + 6736200d = 9068700$$

Step Three: Let us solve the above system of four simultaneous algebraic equations for the four constants a, b, c, and d using the inverse matrix method (check Chapter 10 on Solving Equations for details):

```
>> A = [n1 sumx sumy sumz ; sumx
sumxx sumxy sumxz ; sumy sumxy
sumyy sumyz ; sumz sumxz sumyz
sumzz]

A =

           7         1825           53
6810
        1825       478531        14284
1781580
          53        14284          489
53390
        6810      1781580        53390
6736300

>> b = [sumw ; sumxw ; sumyw ;
sumzw]

b =

        9190
     2417850
       74000
     9068700

>> inv(A)*b
```

```
ans =

 -121.7772
    2.4977
   24.2467
    0.6166

>> a = ans(1)

a =

 -121.7772

>> b = ans(2)

b =

    2.4977

>> c = ans(3)

c =

   24.2467

>> d = ans(4)

d =

    0.6166
```

Step Four: Let us write the equation of the best fitting hyperplane as follows:

$$w = -121.7772 + 2.4977x + 24.2467y + 0.6166z$$

Step Five: Let us calcuale the approximate values for w to obtained from the best fitting hyperplane equation above:

```
>> w1 = a + b*x + c*y + d*z

w1 =

   1.0e+03 *

    1.0600          1.0866          1.2510
1.3258          1.3506          1.5592
1.5568
```

It is seen from the above results that the approximate values obtained through the best fit are very close to the actual data points given in the Exercise. In this case, there is no need to generate a graph because MATLAB does not have a `plot4` command. Also, there is no need to calculate the errors.

3. Consider a set of n data points (x_i, y_i, z_i, w_i) in four dimensions. In this case and since we are dealing with four-dimensional space, we obtain a best fitting hyperplane instead of a best fitting line. The equation of the best fitting hyperplane can be written in general as follows:

$$w = a + bx + cy + dz$$

What is required is find the values of the four constants a, b, c. and d The same procedure of Chapter 15 is followed here but note that we are dealing with four dimensions instead of three dimensions. The above equation is repeated below with subscript $i = 1, 2, 3, \ldots\ldots n$:

$$w_i = a + bx_i + cy_i + dz_i$$

(E*)

It should be noted that the above equations is repeated n times, where n is the number of data points. Let us write the above equations for the first few data points as $i = 1, 2, 3, \ldots\ldots n$:

$$w_1 = a + b x_1 + c y_1 + d z_1$$

$$w_2 = a + b x_2 + c y_2 + d z_2$$

$$w_3 = a + b x_3 + c y_3 + d z_3$$

$$\ldots\ldots\ldots\ldots\ldots$$

$$\ldots\ldots\ldots\ldots\ldots$$

$$\ldots\ldots\ldots\ldots\ldots$$

$$w_n = a + b x_n + c y_n + d z_n$$

Let us now add the above list of equations to obtain the following expression:

$$\begin{aligned} w_1 + w_2 + w_3 + \ldots\ldots + w_n = {} & a + b x_1 + c y_1 + d z_1 + a + b x_2 + c y_2 + d z_2 \\ & + a + b x_3 + c y_3 + d z_3 \ldots\ldots + a + b x_n + c y_n + d z_n \end{aligned}$$

The above equation can be written in the following simplified form:

$$w_1 + w_2 + w_3 + \ldots\ldots + w_n = (a + a + a + \ldots. + a)$$
$$+ (b\,x_1 + b\,x_2 + b\,x_3 + \ldots\ldots + b\,x_n) + (c\,y_1 + c\,y_2 + c\,y_3 + \ldots\ldots + c\,y_n)$$
$$+ (d\,z_1 + d\,z_2 + d\,z_3 + \ldots\ldots + d\,z_n)$$

where the sum of the a's in the first parenthesis is n times. The above equation can be written using the summation sign $\sum$ as follows:

$$\sum_{i=1}^{n} w_i = n\,a + b\sum_{i=1}^{n} x_i + c\sum_{i=1}^{n} y_i + d\sum_{i=1}^{n} z_i$$
(F*)

where the following summation notation is used in equation (F*):

$$\sum_{i=1}^{n} w_i = w_1 + w_2 + w_3 + \ldots\ldots.. + w_n$$

$$n\,a = a + a + a + \ldots.. + a \qquad (summed\ \ n\ \ times)$$

$$b\sum_{i=1}^{n} x_i = b\,(x_1 + x_2 + x_3 + \ldots\ldots + x_n) = b\,x_1 + b\,x_2 + b\,x_3 + \ldots. + b\,x_n$$

$$c\sum_{i=1}^{n} y_i = c\,(y_1 + y_2 + y_3 + \ldots\ldots + y_n) = c\,y_1 + c\,y_2 + c\,y_3 + \ldots. + c\,y_n$$

$$d\sum_{i=1}^{n} z_i = d(z_1 + z_2 + z_3 + ... + z_n) = d\,z_1 + d\,z_2 + d\,z_3 + ... + d\,z_n$$

The next step would be to multiply equation (E*) by x_i to obtain the following equation:

$$x_i\,w_i = a\,x_i + b\,x_i\,x_i + c\,x_i\,y_i + d\,x_i\,z_i$$
(G*)

Writing the above equation out for each value $i = 1, 2, 3, ...n$, we obtain:

$$x_1\,w_1 = a\,x_1 + b\,x_1^2 + c\,x_1\,y_1 + d\,x_1\,z_1$$

$$x_2\,w_2 = a\,x_2 + b\,x_2^2 + c\,x_2\,y_2 + d\,x_2\,z_2$$

$$x_3\,w_3 = a\,x_3 + b\,x_3^2 + c\,x_3\,y_3 + d\,x_3\,z_3$$

....

....

....

$$x_n\,w_n = a\,x_n + b\,x_n^2 + c\,x_n\,y_n + d\,x_n\,z_n$$

Adding the above list of equations results in the following expression:

$$x_1\,w_1 + x_2\,w_2 + x_3\,w_3 + ... + x_n\,w_n = a\,x_1 + b\,x_1^2 + c\,x_1 y_1 + d\,x_1 z_1$$
$$+ a\,x_2 + b\,x_2^2 + c\,x_2\,y_2 + d\,x_2\,z_2 + a\,x_3 + b\,x_3^2 + c\,x_3\,y_3 + d\,x_3\,z_3$$
$$+ .... + a\,x_n + b\,x_n^2 + c\,x_n y_n + d\,x_n z_n$$

The above equation can be simplified as follows:

$$\begin{aligned} x_1 w_1 + x_2 w_2 + x_3 w_3 + \ldots\ldots + x_n w_n &= a x_1 + a x_2 + a x_3 + \ldots\ldots + a x_n \\ &+ b x_1^2 + b x_2^2 + b x_3^2 + \ldots\ldots + b x_n^2 \\ &+ c x_1 y_1 + c x_2 y_2 + c x_3 y_3 + \ldots\ldots + c x_n y_n \\ &+ d x_1 z_1 + d x_2 z_2 + d x_3 z_3 + \ldots\ldots + d x_n z_n \end{aligned}$$

The above equation can be written using the summation sign $\sum$ as follows:

$$\sum_{i=1}^{n} x_i w_i = a \sum_{i=1}^{n} x_i + b \sum_{i=1}^{n} x_i^2 + c \sum_{i=1}^{n} x_i y_i + d \sum_{i=1}^{n} x_i z_i$$

(H*)

where the following summation notation is used in equation (H*):

$$\sum_{i=1}^{n} x_i w_i = x_1 w_1 + x_2 w_2 + x_3 w_3 + \ldots\ldots + x_n w_n$$

$$a \sum_{i=1}^{n} x_i = a(x_1 + x_2 + x_3 + \ldots\ldots + x_n) = a x_1 + a x_2 + a x_3 + \ldots\ldots + a x_n$$

$$b \sum_{i=1}^{n} x_i^2 = b(x_1^2 + x_2^2 + x_3^2 + \ldots\ldots + x_n^2) = b x_1^2 + b x_2^2 + b x_3^2 + \ldots\ldots + b x_n^2$$

$$c\sum_{i=1}^{n} x_i\, y_i = c(x_1\, y_1 + x_2\, y_2 + x_3\, y_3 + \ldots\ldots + x_n\, y_n)$$

$$= c\,x_1\, y_1 + c\,x_2\, y_2 + c\,x_3\, y_3 + \ldots\ldots + c\,x_n\, y_n$$

$$d\sum_{i=1}^{n} x_i\, z_i = d(x_1\, z_1 + x_2\, z_2 + x_3\, z_3 + \ldots\ldots + x_n\, z_n)$$

$$= d\,x_1\, z_1 + d\,x_2\, z_2 + d\,x_3\, z_3 + \ldots\ldots + d\,x_n\, z_n$$

The next step would be to multiply equation (E*) by y_i to obtain the following equation:

$$y_i\, w_i = a\, y_i + b\, x_i\, y_i + c\, y_i\, y_i + d\, y_i\, z_i$$
(J*)

Writing the above equation out for each value $i = 1, 2, 3, \ldots\ldots n$, we obtain:

$$y_1\, w_1 = a\, y_1 + b\, x_1\, y_1 + c\, y_1^2 + d\, y_1\, z_1$$

$$y_2\, w_2 = a\, y_2 + b\, x_2\, y_2 + c\, y_2^2 + d\, y_2\, z_2$$

$$y_3\, w_3 = a\, y_3 + b\, x_3\, y_3 + c\, y_3^2 + d\, y_3\, z_3$$

....

....

....

$$y_n\, w_n = a\, y_n + b\, x_n\, y_n + c\, y_n^2 + d\, y_n\, z_n$$

Adding the above list of equations results in the following expression:

$$y_1 w_1 + y_2 w_2 + y_3 w_3 + ... + y_n w_n = a y_1 + b x_1 y_1 + c y_1^2 + d y_1 z_1$$
$$+ a y_2 + b x_2 y_2 + c y_2^2 + d y_2 z_2 + a x_3 + b x_3 y_3 + c y_3^2 + d y_3 z_3$$
$$+ .... + a y_n + b x_n y_n + c y_n^2 + d y_n z_n$$

The above equation can be simplified as follows:

$$y_1 w_1 + y_2 w_2 + y_3 w_3 + ... + y_n w_n = a y_1 + a y_2 + a y_3 + ... + a y_n$$
$$+ b x_1 y_1 + b x_2 y_2 + b x_3 y_3 + ... + b x_n y_n$$
$$+ c y_1^2 + c y_2^2 + c y_3^2 + ... + c y_n^2$$
$$+ d y_1 z_1 + d y_2 z_2 + d y_3 z_3 + ... + d y_n z_n$$

The above equation can be written using the summation sign $\sum$ as follows:

$$\sum_{i=1}^{n} y_i w_i = a \sum_{i=1}^{n} y_i + b \sum_{i=1}^{n} x_i y_i + c \sum_{i=1}^{n} y_i^2 + d \sum_{i=1}^{n} y_i z_i$$

(K*)

where the following summation notation is used in equation (H):

$$\sum_{i=1}^{n} y_i w_i = y_1 w_1 + y_2 w_2 + y_3 w_3 + ... + y_n w_n$$

$$a \sum_{i=1}^{n} y_i = a(y_1 + y_2 + y_3 + ... + y_n) = a y_1 + a y_2 + a y_3 + ... + a y_n$$

$$b\sum_{i=1}^{n} x_i\, y_i = b(x_1\, y_1 + x_2\, y_2 + x_3\, y_3 + ... + x_n\, y_n)$$

$$= b\, x_1\, y_1 + b\, x_2\, y_2 + b\, x_3\, y_3 + ... + b\, x_n\, y_n$$

$$c\sum_{i=1}^{n} y_i^2 = c(y_1^2 + y_2^2 + y_3^2 + ... + y_n^2) = c\, y_1^2 + c\, y_2^2 + c\, y_3^2 + ... + c\, y_n^2$$

$$d\sum_{i=1}^{n} y_i\, z_i = d(y_1\, z_1 + y_2\, z_2 + y_3\, z_3 + ... + y_n\, z_n)$$

$$= d\, y_1\, z_1 + d\, y_2\, z_2 + d\, y_3\, z_3 + ... + d\, y_n\, z_n$$

The next step would be to multiply equation (E*) by z_i to obtain the following equation:

$$z_i\, w_i = a\, z_i + b\, x_i\, z_i + c\, y_i\, z_i + d\, z_i^2$$

(L*)

Writing the above equation out for each value $i = 1, 2, 3, ...n$, we obtain:

$$z_1\, w_1 = a\, z_1 + b\, x_1\, z_1 + c\, y_1\, z_1 + d\, z_1^2$$

$$z_2\, w_2 = a\, z_2 + b\, x_2\, z_2 + c\, y_2\, z_2 + d\, z_2^2$$

$$z_3\, w_3 = a\, z_3 + b\, x_3\, z_3 + c\, y_3\, z_3 + d\, z_3^2$$

....

....

....

$$z_n w_n = a z_n + b x_n z_n + c y_n z_n + d z_n^2$$

Adding the above list of equations results in the following expression:

$$\begin{aligned} z_1 w_1 + z_2 w_2 + z_3 w_3 + \ldots\ldots + z_n w_n &= a z_1 + b x_1 z_1 + d y_1 z_1 + d z_1^2 + \\ &+ a z_2 + b x_2 z_2 + c y_2 z_2 + d z_2^2 + a z_3 + b x_3 z_3 + c y_3 z_3 + d z_3^2 + \\ &+ \ldots\ldots + a z_n + b x_n z_n + c y_n z_n + d z_n^2 \end{aligned}$$

The above equation can be simplified as follows:

$$\begin{aligned} z_1 w_1 + z_2 w_2 + z_3 w_3 + \ldots\ldots + z_n w_n &= a z_1 + a z_2 + a z_3 + \ldots\ldots + a z_n \\ &+ b x_1 z_1 + b x_2 z_2 + b x_3 z_3 + \ldots\ldots + b x_n z_n \\ &+ c y_1 z_1 + c y_2 z_2 + c y_3 z_3 + \ldots\ldots + c y_n z_n \\ &+ d z_1^2 + d z_2^2 + d z_3^2 + \ldots\ldots + d z_n^2 \end{aligned}$$

The above equation can be written using the summation sign $\sum$ as follows:

$$\sum_{i=1}^{n} z_i w_i = a \sum_{i=1}^{n} z_i + b \sum_{i=1}^{n} x_i z_i + c \sum_{i=1}^{n} y_i z_i + d \sum_{i=1}^{n} z_i^2$$

(M*)

where the following summation notation is used in equation (M*):

$$\sum_{i=1}^{n} z_i w_i = z_1 w_1 + z_2 w_2 + z_3 w_3 + \ldots\ldots + z_n w_n$$

$$a\sum_{i=1}^{n} z_i = a(z_1 + z_2 + z_3 + \ldots\ldots + z_n) = a\,z_1 + a\,z_2 + a\,z_3 + \ldots\ldots + a\,z_n$$

$$b\sum_{i=1}^{n} x_i\, z_i = b(x_1\, z_1 + x_2\, z_2 + x_3\, z_3 + \ldots\ldots + x_n\, z_n)$$

$$= b\,x_1\, z_1 + b\,x_2\, z_2 + b\,x_3\, z_3 + \ldots\ldots + b\,x_n\, z_n$$

$$c\sum_{i=1}^{n} y_i\, z_i = c(y_1\, z_1 + y_2\, z_2 + y_3\, z_3 + \ldots\ldots + y_n\, z_n)$$

$$= c\,y_1\, z_1 + c\,y_2\, z_2 + c\,y_3\, z_3 + \ldots\ldots + c\,y_n\, z_n$$

$$d\sum_{i=1}^{n} z_i^2 = d(z_1^2 + z_2^2 + z_3^2 + \ldots\ldots + z_n^2) = d\,z_1^2 + d\,z_2^2 + d\,z_3^2 + \ldots\ldots + d\,z_n^2$$

Next we list equations (F*), (H*), (K*), and (M*) as shown in the boxed list below. These four equations should be solved simultaneously to obtain the four constants a, b, c, and d.

$$\sum_{i=1}^{n} w_i = na + b\sum_{i=1}^{n} x_i + c\sum_{i=1}^{n} y_i + d\sum_{i=1}^{n} z_i$$

(F*)

$$\sum_{i=1}^{n} x_i w_i = a\sum_{i=1}^{n} x_i + b\sum_{i=1}^{n} x_i^2 + c\sum_{i=1}^{n} x_i y_i + d\sum_{i=1}^{n} x_i z_i$$

(H*)

$$\sum_{i=1}^{n} y_i w_i = a\sum_{i=1}^{n} y_i + b\sum_{i=1}^{n} x_i y_i + c\sum_{i=1}^{n} y_i^2 + d\sum_{i=1}^{n} y_i z_i$$

(K*)

$$\sum_{i=1}^{n} z_i w_i = a\sum_{i=1}^{n} z_i + b\sum_{i=1}^{n} x_i z_i + c\sum_{i=1}^{n} y_i z_i + d\sum_{i=1}^{n} z_i^2$$

(M*)

References

MATLAB Tutorials

1. http://www.cyclismo.org/tutorial/matlab/

   This is an online tutorial in HTML format from Union College, New York. The tutorial emphasizes vectors, matrices, vector operations, loops, plots, executable files (scripts), subroutines (functions), if statements, and data files.

2. http://www.math.ufl.edu/help/matlab-tutorial/

   This is an online tutorial in HTML format from the University of Florida, Gainesville. The tutorial emphasizes matrices, variables, functions, decisions, loops, and scripts. This tutorial has a nice list of important commands of MATLAB with a short description of each command.

3. http://www.mines.utah.edu/gg_computer_seminar/matlab/matlab.html

   This is an online tutorial in HTML format available at the website of University of Utah, Salt Lake City (Original by Kermit Sigmon, University of Florida). This tutorial emphasizes matrices, decisions, loops, scalar functions, vector functions, matrix functions, strings, and graphics. You may also download the tutorial as a postscript file (39 pages that include a comprehensive reference section).

4. http://web.mit.edu/afs/.athena/astaff/project/logos/olh/Math/Matlab/Matlab.html

   This is an online tutorial in HTML format hosted at MIT. It emphasizes matrices, arithmetic and logical operators, control structures, selective indexing, polynomial operations, signal processing functions, graphics, scripts, and functions.

5. http://www.mit.edu/people/abbe/matlab/main.html

   This is an online 3-day tutorial at MIT in HTML format. It covers matrices, various MATLAB commands, help, plotting, polynomials, variables, scripts, functions, loops, debugging, differential equations, vectorization, three-dimensional graphics, and symbolic math.

6. http://texas.math.ttu.edu/~gilliam/ttu/m4330/m4330.html

   This is a tutorial hosted at Texas Tech University, Lubbock. They provide six PDF files for download in the form of six lessons for a mathematics course they teach there. These lessons are somewhat advanced and specialized and are not recommended for beginner students of MATLAB.

7. http://www.engin.umich.edu/group/ctm/basic/basic.html

   This is a basic tutorial from Carnegie Mellon and the University of Michigan. It covers the basics of vectors, functions, plotting, polynomials, matrices, printing, help, and M-files.

8. http://www.mathworks.com/academia/student_center/tutorials/launchpad.html

   This is an online tutorial from the MathWorks, the company that develops and sells MATLAB. This tutorial covers variables, calculations, plotting, scripts, and files.

9. http://www.math.utah.edu/lab/ms/matlab/matlab.html#starting

   This is an online short tutorial from the University of Utah, Salt Lake City. This is a very simple tutorial recommended for beginners. It covers matrices, vectors, systems of equations, loops, and graphing in one and two dimensions.

10. http://www.math.mtu.edu/~msgocken/intro/intro.html

    This is an online tutorial in HTML format hosted at Michigan Tech and written by Mark S. Gockenback. It covers calculations, graphs, programming, advanced matrix calculations, advanced graphics, solving nonlinear problems, and advanced data types like structures, cell arrays, and objects.

11. http://www.eece.maine.edu/mm/matweb.html

    MATLAB Educational Websites: This is an online HTML page containing links to several educational websites on various topics using MATLAB. The page contains also links to numerous MATLAB tutorials.

12. http://web.cecs.pdx.edu/~gerry/MATLAB/

    MATLAB Hypertext Reference: This is an online tutorial in HTML format written by Gerald Recktenwald and hosted at Portland State University, Portland. This tutorial covers variables, plotting, and programming.

13. http://www.facstaff.bucknell.edu/maneval/help211/helpmain.html

    Helpful Information for Using MATLAB: These pages of helpful information are maintained by Jim Maneval and hosted at Bucknell University, Lewisburg.

14. http://www.indiana.edu/~statmath/math/matlab/gettingstarted/index.html

    Getting Started with MATLAB: This is an online tutorial in HTML format. The tutorial may also be downloaded and printed as a PDF file (14 pages). This tutorial covers syntax, matrices, graphics, and programming.

15. www.math.udel.edu/~driscoll/teaching/**matlab**_adv.pdf

    Crash Course in MATLAB: This tutorial is in the form of a PDF file download. It is written by Tobin A. Driscoll and hosted at the University of Delaware, The tutorial is 66 pages that can be downloaded and printed. This tutorial covers arrays, matrices, scripts, functions, errors, 2D and 3D graphics, color, handles and properties, vectorization, advanced data structures (strings, cell arrays, structures), linear algebra, optimization, data fitting, quadrature, and differential equations.

16. www.maths.dundee.ac.uk/~ftp/na-reports/**MatlabNotes**.pdf

    An Introduction to MATLAB: This tutorial is in the form of a PDF file download. It is written by David F. Griffiths and is hosted at the University of Dundee. The tutorial is 37 pages that can be downloaded and printed. This tutorial covers numbers, formats, variables, output, vectors, 2D plotting, scripts, vector operations, matrices, matrix operations, systems of linear equations, strings, loops, decisions, 3D plotting, files, and graphical user interfaces.

17. www.geosci.uchicago.edu/~gidon/geosci236/organize/**matlab**Intro.pdf

    An Introduction to MATLAB: This tutorial is in the form of a PDF file download. It is written by S. Butenko, P. Pardalos, and L. Pitsoulis. It is hosted at the University of Chicago. The tutorial is 28 pages that can be downloaded and printed. This tutorial covers matrices, matrix operations, workspace, formats, functions, vector functions, matrix functions, polynomial functions, programming, loops, decisions, M-files, and 2D and 3D graphics.

18. http://www.math.colostate.edu/~gerhard/classes/340/notes/index.html

    MATLAB Notes: This tutorial is in the form HTML pages online. It is hosted at Colorado State University. This tutorial covers arithmetic, array operations, scripts, functions, plotting, polynomials, matrices, vectors, linear systems of algebraic equations, and symbolic computation. There are some links to more tutorials on the internet that are hosted at other places.

19. http://www.indiana.edu/~statmath/math/matlab/gettingstarted/index.html

    Getting Started with MATLAB: This tutorial is in the form of online HTML pages that are hosted at Indiana University. It may also be downloaded as a PDF file (14 pages) and printed. This tutorial covers syntax, matrices, graphics, and programming.

20. http://www-ccs.ucsd.edu/matlab/toolbox/symbolic/symbmath.html

This is an online reference guide for the MATLAB Symbolic Math Toolbox. It is hosted at the University of California, San Diego. The tutorial covers about 85 MATLAB symbolic math commands that are listed alphabetically with examples.

21. http://www.phys.ufl.edu/docs/matlab/toolbox/symbolic/ch1.html

    Using the Symbolic Math Toolbox: This tutorial is in the form of HTML pages that is hosted at the University of Florida, Gainesville. This tutorial seems to be comprehensive and covers calculus, simplifications, substitutions, variable-precision arithmetic, linear algebra, solving equations, special mathematical functions, using Maple functions, and the extended symbolic math toolbox.

22. http://en.wikipedia.org/wiki/Matlab

    This is a small overview in the form of an HTML page hosted at the website of Wikipedia, the free internet encyclopedia. The page has a brief history of MATLAB and some useful links.

23. http://www.facstaff.bucknell.edu/maneval/help211/exercises.html

    MATLAB Exercises: This web page provides some MATLAB exercises for practice. The site is hosted at Bucknell University. The exercises cover syntax, arrays, relational operators, logical operators, decisions, loops, and programming.

24. http://www.math.chalmers.se/~nilss/enm/INTRO_matlab_I.html

    MATLAB Exercises: This web page provides three sets of MATLAB exercises for practice. The site is hosted at Chalmers University of Technology and Gteborg University. The exercises cover arithmetic, formats, variables, vectors, matrices, decisions, scripts, loops, computer arithmetic, strings, and graphics.

25. http://college.hmco.com/mathematics/larson/elementary_linear/5e/students/matlabs.html

    MATLAB Exercises: These are MATLAB exercises taken from the book "Elementary Linear Algebra", fifth edition, by Ron Larson. Each set of exercises is downloaded as a PDF file that can be printed. These exercises cover systems of linear equations, matrices, determinants, vector spaces, inner product spaces, linear transformations, eigenvalues and eigenvectors, and complex vector spaces.

26. http://www.math.umn.edu/~ROberts/math5385/matlabEx1.html

    Supplementary MATLAB Exercises: These exercises are hosted at the University of Minnesota. They are written specifically for a mathematics course at the university. These exercises mostly deal with the graphics capabilities of MATLAB.

27. www.kom.aau.dk/~borre/**matlab**7/exercise.pdf

    This is a PDF file download of a set of 8 pages of MATLAB exercises. The exercises are somewhat advanced and are not recommended for beginning students of MATLAB.

28. www.chaos.swarthmore.edu/courses/Phys50L_2006/**Matlab**/**MatlabExercises**.pdf

    This is a PDF file download of 27 pages of MATLAB exercises. These exercises are written for a physics course and are somewhat advanced. They are not recommended for beginners.

MATLAB Books:

29. Gilat, A., *MATLAB: An Introduction with Applications*, Second Edition, John Wiley & Sons, 2004.

30. Pratap, R., *Getting Started with MATLAB 7: An Introduction for Scientists and Engineers*, Oxford University Press, 2005.

31. Hanselmann, D. and Littlefield, B., *Mastering MATLAB 7*, Prentice Hall, 2004.

32. Palm, W., *Introduction to MATLAB 7 for Engineers*, McGraw-Hill, 2004.

33. Moore, H., *MATLAB for Engineers*, Prentice Hall, 2006.

34. Chapman, S., *MATLAB Programming for Engineers*, Thomson Engineering, 2004.

35. Davis, T. and Sigmon, K., *MATLAB Primer*, Seventh Edition, Chapman & Hall, 2004.

36. Higham, D. and Higham, N., *MATLAB Guide*, Second Edition, SIAM, 2005.

37. King, J., *MATLAB for Engineers*, Addison-Wesley, 1988.

38. Etter, D., *Introduction to MATLAB for Engineers and Scientists*, Prentice Hall, 1995.

39. Magrab, E. et al., *An Engineer's Guide to MATLAB*, Prentice Hall, 2000.

40. Etter, D., Kuncicky, D. and Hull, D., *Introduction to MATLAB 6*, Prentice Hall, 2001.

41. Recktenwald, G., *Introduction to Numerical Methods and MATLAB: Implementation and Applications*, Prentice Hall, 2001.

42. Biran, A. and Breiner, M., *MATLAB 5 for Engineers*, Addison-Wesley, 1999.

43. Part-Enander, E. and Sjoberg, A., *The MATLAB 5 Handbook*, Addison-Wesley, 1999.

44. Etter, D., *Engineering Problem Solving with MATLAB*, Prentice Hall, 1993.

45. Chen, K., Giblin, P, and Irving, A., *Mathematical Explorations with MATLAB*, Cambridge University Press, 1999.

46. Mathews, J. and Fink, K., *Numerical Methods Using MATLAB*, Third Edition, Prentice Hall, 1999.

47. Fausett, L., *Applied Numerical Analysis Using MATLAB*, Prentice Hall, 1999.

Statistics Books:

48. Harper, W. M., *Statistics*, Fifth Edition, Longman Group UK, 1988.

49. Daniel, W. W. and Terrell, J. C., *Business Statistics: Basic Concepts and Methodology*, Third Edition, Houghton Mifflin Company, 1983.

50. Spiegel, M. R., *Theory and Problems of Statistics: Schaum's Outline Series*, McGraw-Hill, 1972.

51. Swinscow, T. D. V., *Statistics at Square One*, The British Medical Journal, 1991.

52. Freedman, D., Pisani, R., Purves, R., and Adhikari, A., *Statistics*, Second Edition, W. W. Norton & Company, 1991.

53. Wesley Barnes, J., *Statistical Analysis for Engineers and Scientists:: A Computer-Based Approach*, McGraw-Hill, 1994.

54. Bhattacharyya, G. K. and Johnson, R. A., *Statistical Concepts and Methods*, Wiley, 1077.

55. Steel, R. G. D. and Torrie, J. H., *Principles and Procedures of Statistics: A Biometric Approach*, McGraw-Hill, 1980.

56. Rumsey, D. J., *Statistics for Dummies*, Second Edition, For Dummies Series, 2011.

57. Urdan, T. C., *Statistics in Plain English*, Third Edition, Routledge, 2010.

58. White, R. S. and White, J. S., *Statistics*, Tenth Edition, Wiley, 2013.

59. McClave, J. T. and Sincich, T. T., *Statistics*, Twelfth Edition, Pearson, 2012.

60. Triola, M. F., *Elementary Statistics*, Twelfth Edition, Pearson, 2012.

Statistics/MATLAB Books:

61. Matinez, W. L. and Marinez, A. R., *Computational Statistics Handbook with MATLAB*, Third Edition, Chapman & Hall - CRC, 2015.

62. Cho, M. and Martinez, W. L., *Statistics in MATLAB: A Primer*, Chapman & Hall – CRC, 2014.

63. Lopez, C. P., *Statistics with MATLAB: Examples and Exercises*, Createspace Independent Publishing Platform, 2013.

64. Martinez, W. L. and Martinez, A. R., *Exploratory Data Analysis with MATLAB*, Second Edition, Chapman & Hall – CRC, 2010.

65. Braselton, J., *Data Analysis with MATLAB*, Createspace Independent Publishing Platform, 2016.

Regression Analysis Books:

66. Schroeder, L. D., Sjoquist, D. L., and Stepham P. E., *Understanding Regression Analysis*, SAGE Publications, 1986.

67. Montgomery, D. C. and Peck, E. A., *Introduction to Linear Regression Analysis*, Fifth Edition, Wiley, 2012.

68. Mendenhall, W. and Sincich, T. T., *A Second Course in Statistics: Regression Analysis*, Seventh Edition, Pearson, 2011.

69. Draper, N. R. and Smith, H., *Applied Regression Analysis*, Wiley-Interscience, 1998.

70. Chatterjee, S. and Hadi, A. S., *Regression Analysis by Example*, Fifth Edition, Wiley, 2012.

71. Sen, A. and Srivastava, M., *Regression Analysis: Theory, Methods, and Applications*, Springer, 1997.

72. Hoffmann, J. P. and Shafer, K., *Linear Regression Analysis: Assumptions and Applications*, NASW Press, 2015.

Regression Analysis/MATLAB Books:

73. Braselton, J., *Curve Fitting with MATLAB: Linear and Nonlinear Regression, Interpolation*, Createspace Independent Publishing Platform, 2016.

74. Matinez, W. L. and Marinez, A. R., *Computational Statistics Handbook with MATLAB*, Third Edition, Chapman & Hall - CRC, 2015.

Installation of MATLAB

In this book, it is assumed that you have already installed MATLAB on your computer system. To go through the eleven chapters of the book, you need to have MATLAB running on your computer. For help on installing MATLAB on your computer, check the following web links:

MATLAB Installation for Windows:

http://www.itc.virginia.edu/research/matlab/install/installwin.html

MATLAB Installation for Linux:

http://www.freebsd.org/doc/zh_TW/books/handbook/linuxemu-matlab.html

MATLAB Installation for Macintosh:

http://www.itc.virginia.edu/research/matlab/install/installmac.html

MATLAB Installation for Other Systems:

http://shum.huji.ac.il/cc/matlR13linuxhomeinst.html

Index

www.ingramcontent.com/pod-product-compliance
Lightning Source LLC
LaVergne TN
LVHW090406210926
848480LV00002B/471
* 9 7 9 8 2 9 5 5 4 9 1 4 4 *